NEUROMETHODS ☐ 28

Transgenic *Xenopus*

NEUROMETHODS

Series Editors: Alan A. Boulton and Glen B. Baker

28. **Transgenic *Xenopus*:** *Microinjection Methods and Developmental Neurobiology*, by ***Shlomo Seidman and Hermona Soreq***, 1997
27. **Voltammetric Methods in Brain Systems**, edited by ***Alan A. Boulton, Glen B. Baker, and Ralph N. Adams***, 1995
26. **Patch-Clamp Applications and Protocols**, edited by ***Alan A. Boulton, Glen B. Baker, and Wolfgang Walz***, 1995
25. **Neurotrophic Factors**, edited by ***Alan A. Boulton, Glen B. Baker, and Franz Hefti***, 1993
24. **Animal Models of Drug Addiction**, edited by ***Alan A. Boulton, Glen B. Baker, and Peter Wu***, 1992
23. **Practical Cell Culture Techniques**, edited by ***Alan A. Boulton, Glen B. Baker, and Wolfgang Walz***, 1992
22. **Animal Models of Neurological Disease, II:** *Metabolic Encephalopathies and the Epilepsies*, edited by ***Alan A. Boulton, Glen B. Baker, and Roger F. Butterworth***, 1992
21. **Animal Models of Neurological Disease, I:** *Neurodegenerative Disease*, edited by ***Alan A. Boulton, Glen B. Baker, and Roger F. Butterworth***, 1992
20. **Intracellular Messengers**, edited by ***Alan A. Boulton, Glen B. Baker, and Colin W. Taylor***, 1992
19. **Animal Models in Psychiatry, II**, edited by ***Alan A. Boulton, Glen B. Baker, and Mathew T. Martin-Iverson***, 1991
18. **Animal Models in Psychiatry, I**, edited by ***Alan A. Boulton, Glen B. Baker, and Mathew T. Martin-Iverson***, 1991
17. **Neuropsychology**, edited by ***Alan A. Boulton, Glen B. Baker, and Merrill Hiscock***, 1990
16. **Molecular Neurobiological Techniques**, edited by ***Alan A. Boulton, Glen B. Baker, and Anthony T. Campagnoni***, 1990
15. **Neurophysiological Techniques:** *Applications to Neural Systems*, edited by ***Alan A. Boulton, Glen B. Baker, and Case H. Vanderwolf***, 1990
14. **Neurophysiological Techniques:** *Basic Methods and Concepts*, edited by ***Alan A. Boulton, Glen B. Baker, and Case H. Vanderwolf***, 1990
13. **Psychopharmacology**, edited by ***Alan A. Boulton, Glen B. Baker, and Andrew J. Greenshaw***, 1989
12. **Drugs as Tools in Neurotransmitter Research**, edited by ***Alan A. Boulton, Glen B. Baker, and Augusto V. Juorio***, 1989
11. **Carbohydrates and Energy Metabolism**, edited by ***Alan A. Boulton, Glen B. Baker, and Roger F. Butterworth***, 1989
10. **Analysis of Psychiatric Drugs**, edited by ***Alan A. Boulton, Glen B. Baker, and Ronald T. Coutts***, 1988
9. **The Neuronal Microenvironment**, edited by ***Alan A. Boulton, Glen B. Baker, and Wolfgang Walz***, 1988
8. **Imaging and Correlative Physicochemical Techniques**, edited by ***Alan A. Boulton, Glen B. Baker, and Donald P. Boisvert***, 1988
7. **Lipids and Related Compounds**, edited by ***Alan A. Boulton, Glen B. Baker, and Lloyd A. Horrocks***, 1988
6. **Peptides**, edited by ***Alan A. Boulton, Glen B. Baker, and Quentin Pittman***, 1987
5. **Neurotransmitter Enzymes**, edited by ***Alan A. Boulton, Glen B. Baker, and Peter H. Yu***, 1986
4. **Receptor Binding Techniques**, edited by ***Alan A. Boulton, Glen B. Baker, and Pavel D. Hrdina***, 1986
3. **Amino Acids**, edited by ***Alan A. Boulton, Glen B. Baker, and James D. Wood***, 1985
2. **Amines and Their Metabolites**, edited by ***Alan A. Boulton, Glen B. Baker, and Judith M. Baker***, 1985
1. **General Neurochemical Techniques**, edited by ***Alan A. Boulton and Glen B. Baker***, 1985

NEUROMETHODS □ 28

Transgenic *Xenopus*

Microinjection Methods and Developmental Neurobiology

by

Shlomo Seidman
and Hermona Soreq
The Hebrew University of Jerusalem
Jerusalem, Israel

Humana Press Totowa, New Jersey

Softcover reprint of the hardcover 1st edition 1997
999 Riverview Drive, Suite 208
Totowa, New Jersey 07512

This publication is printed on acid-free paper. ∞
ANSI Z39.48-1984 (American National Standards Institute)
Permanence of Paper for Printed Library Materials.

For additional copies, pricing for bulk purchases, and/or information about other Humana titles, contact Humana at the above address or at any of the following numbers:
Tel.: 201-256-1699; Fax: 201-256-8341; E-mail: humana@interramp.com

Cover illustration: Two-day-old *Xenopus* embryo microinjected with plasmid DNA encoding a variant form of human acetylcholinesterase and stained for catalytically active enzyme. Photograph was taken with a Zeiss Stemi SV11 stereomicroscope fitted with an MC 80 camera. Illustration courtesy of Meira Sternfeld.

Cover design by Patricia F. Cleary.

ISBN 978-1-61737-045-8 e-ISBN 978-1-59259-633-1
ISSN 0893-2336

Preface to the Series

When the President of Humana Press first suggested that a series on methods in the neurosciences might be useful, one of us (AAB) was quite skeptical; only after discussions with GBB and some searching both of memory and library shelves did it seem that perhaps the publisher was right. Although some excellent methods books had recently appeared, notably in neuroanatomy, it was a fact that there was a dearth in this particular field, a fact attested to by the alacrity and enthusiasm with which most of the contributors to this series accepted our invitations and suggested additional topics and areas. After a somewhat hesitant start, essentially in the neurochemistry section, the series has grown and will encompass neurochemistry, neuropsychiatry, neurology, neuropathology, neurogenetics, neuroethology, molecular neurobiology, animal models of nervous disease, and no doubt many more "neuros." Although we have tried to include adequate methodological detail and in many cases detailed protocols, we have also tried to include wherever possible a short introductory review of the methods and/or related substances, comparisons with other methods, and the relationship of the substances being analyzed to neurological and psychiatric disorders. Recognizing our own limitations, we have invited a guest editor to join with us on most volumes in order to ensure complete coverage of the field. These editors will add their specialized knowledge and competencies. We anticipate that this series will fill a gap; we can only hope that it will be filled appropriately and with the right amount of expertise with respect to each method, substance or group of substances, and area treated.

Alan A. Boulton
Glen B. Baker

Preface

The need to better understand the molecular, biochemical, and cellular processes by which a developing neuronal system unfolds has led to the development of a unique set of experimental tools and organisms. Special emphasis was devoted to allowing us access, at the earliest stages, to the genomic basis underlying the system's ultimate complexity, as exhibited once its structures are fully formed. Yet, nerve cells are anatomically, physiologically, and biochemically diverse. The multitude of distinctly different routes for their development thus makes the developing nervous system especially intriguing for molecular neurobiologists. In particular, the demands of modern molecular neuroscience call for the establishment of efficient yet versatile systems for studying these complex processes.

Transgenic embryos of the frog *Xenopus laevis* offer an excellent system for approaching neuroscientific issues. Insertion of foreign genes is performed simply, by microinjection under binocular observation; hundreds of in vitro-fertilized embryos can be microinjected in one experiment. Embryos develop in tap water, at room temperature, and within a few days become independent swimming tadpoles with fully functioning neuromuscular systems. Being relatively small, these organisms are amenable to detailed analyses at the levels of mRNA, protein, and cell. Their rapid development permits the study of morphogenetic processes involved in early development, such as myogenesis and neural induction, as well as those involved in organogenesis and formation of the brain, the musculature, and the interconnections between them. Foreign DNA remains predominantly extrachromosomal. Thus, one can use these transiently transgenic animals for asking questions relating to the regulation of gene expression at both the transcriptional and posttranscriptional levels. Moreover, induced changes in development following ectopic overexpression or down-regulation of specific gene products may reveal the role of those proteins in early development.

Though each gene product will necessarily be approached from a different perspective, there are several common thematic questions. The first of these asks: "In what lineages of the developing nervous system will this gene product be expressed?" This question can be addressed by *in situ* hybridization, together with cytochemical or immunochemical means at the levels of optic and electron microscopy. The answer to this question in turn frames the next: "What is the role of this particular gene product in the differentiation of specific neuronal connections?" Another major thematic inquiry is: "What are the rules and the underlying molecular mechanisms that govern synaptogenesis? What are the macromolecules playing roles in this process?" Underlying all of these questions is a concern with developing working hypotheses or explicitly devising testable models that will shape the course of further experimentation. Transgenic *Xenopus* embryos can successfully express genes from other species, splice their mRNA transcripts, and process their polypeptide products in a faithful and efficient manner reproducing the in vivo situation. Amenable for polymerase chain reaction (PCR) and detailed biochemical tests, they can yield answers to long-sought problems in a short time and with limited investment.

Within *Transgenic* Xenopus, we present studies where the human acetylcholinesterase gene was overexpressed in developing neuromuscular junctions of these embryos, enhancing their development in a manner dependent on the muscle-specific pattern of alternative splicing of the corresponding mRNA. Such experiments could not be reproduced in transgenic mice or be the outcome of cell transfection studies; therefore, they exemplify the power of transiently transgenic *Xenopus* embryos in neuroscience research. It is our hope that this volume will be of value to all researchers interested in the molecular, biochemical, and cellular aspects of neuronal development. Toward this end, the text is extensively illustrated and includes detailed experimental protocols and up-to-date reference list.

Shlomo Seidman
Hermona Soreq

Acknowledgments

We express gratitude to our colleagues in the lab (past and present), who have worked together to provide an atmosphere of cordial cooperation within which individual efforts are unselfishly appreciated by others and integrated into coherent group achievements from which all can benefit. In particular, to the following individuals who made direct contributions to the experimental work presented in this book: Revital Ben Aziz-Aloya and Meira Sternfeld, for managing the molecular biology aspects of the project and for sharing fully in the planning, execution, and evaluation of the microinjection experiments; Rina Timberg, for her expert and meticulous management of the electron microscope analyses; Daniella Kaufer-Nachum, for her painstaking and thoughtful contributions to the morphometric analyses; Michael Shapira, for carrying the acetylcholine receptor (AChR) experiments to their conclusion; Dalia Ginzberg, for her tireless commitment to the efficacy and scientific integrity of the group output.

We are grateful to Joel Yisraeli, Hadassah Medical School (Jerusalem), for introducing us to *Xenopus* embryo microinjections; to Richard Harland (Berkeley, CA), for guidance in whole-mount cytochemical staining; and to Hazel Sive and R. Harland, for their instructive CSH course and manual on *Xenopus* development. We would like to thank Baruch Velan and Avigdor Shafferman (Ness-Ziona), for recombinant human AChE (rHAChE) expressed in 293 cells; M. Gorecki (Rehovot), for rHAChE expressed in *Escherichia coli*; Urs Brodbeck and Jian Liao (Switzerland) and Bent Norgaard-Pedersen (Denmark), for their collaborative efforts in the immunochemical work; and Jim Patrick (Houston, TX), for mouse nicotinic AChR (nAChR)-encoding plasmids. We thank David Glick and Christian Andres (Jerusalem) for helpful discussions and Oded Khaner (Jerusalem) for critically reviewing portions of the manuscript. We are grateful to Carl Zeiss Jena GmbH (Germany) for supporting the reproduction of the color plates.

This research was supported by grants to Hermona Soreq from the Israel Basic Research Fund administered by the Israel Academy of Sciences and Humanities, the US Army Medical Research and Development Command (Grant No. DAMD 17-90-Z0038) and the US-Israel Binational Science Foundation (Grant No. 89-00205).

Contents

List of Color Plates

Color plates appear as an insert following p. 110.

Chapter 1

Scientific Background

Xenopus laevis as an Experimental Model System

Xenopus *Development*

The complete reproductive cycle of *Xenopus* spans up to 18 mo from the earliest stages of oogenesis to the sexually mature adult. Owing to its popularity as an embryonic model, the long arduous path to adulthood in *Xenopus* has been thoroughly documented (*see* Deuchar, 1966,1975; Nieuwkoop and Faber, 1967,1994; Dumont, 1972). For an excellent overview of early events in amphibian development—especially *Xenopus*—the reader is referred to Scott Gilbert's textbook *Developmental Biology* (1994). A beautiful color atlas depicting early events in *Xenopus* development is also available (Hausen and Riebesell, 1991). Moreover, a detailed fate map of the 32-cell *Xenopus* embryo has been published (Dale and Slack, 1987). Much of our knowledge about the molecular processes mediating early *Xenopus* development has been obtained using microinjection protocols; some of these achievements are presented in Table 1. A brief description of the principal developmental events pertaining to the use of microinjected *Xenopus* oocytes and embryos are presented in the next section.

Prefertilization

Broken down into 6 stages (I–VI) (Dumont, 1972) (Fig. 1), oogenesis takes about 3 mo. During this time, maternal stores of ribonucleic acid (RNA) (ribosomal RNA [rRNA]

Table 1
Developmental Processes Studied in Microinjected *Xenopus* Embryos

Process	Molecules examined (ref.)[a]
Gastrulation and morphogenetic cell movements	Goosecoid *(1)*; nodal-related *(2)*
Mesoderm induction and dorso-ventral axial patterning	TGF-β receptor *(3)*; Xwnt-8 *(4)*; Raf-1 *(5)*; FGF receptor *(6)*; FGF receptor/activin *(7)*; BMP *(8)*; N-cadherin *(9)*; GSK-3 *(10)*; brachyury/pintallavis *(11)*
Neural induction and antero-posterior axial patterning	Noggin *(12)*; XASH *(13)*; follistatin *(14)*; activin/activinR *(15)*; XIPOU 2 *(16)*; Xotz2 *(17)*; hedgehog *(18)*; Xlim-1 *(19)*
Neuronal differentiation	Pintallavis *(20)*; vhh-1 *(21)*; neuroD *(22)*
Myogenesis	MyoD *(23)*; MyoD *(24)*/Myf5
Somitogenesis	Vimentin/desmin *(25)*
NMJ development	AChE *(26)*; AChR *(27)*; synapsin I *(28)*; synapsin IIa *(29)*; synaptophysin *(30)*
Cell adhesion	EP-cadherin *(31)*; β-catenin *(32)*
Gap junctional communication	Wnt-1 *(33)*; connexins *(34)*
Tissue-specific gene regulation transcriptional	E12 *(35)*/MyoD; distal-less-like-2 (Xdll-2) *(36)*; TFIIIA *(37)*; TFIID *(38)*/MEF2/XMyoD
Tissue-specific gene regulation posttranscriptional	AChE *(39)*

[a]References: *1*, Niehrs et al. (1993); *2*, Smith et al. (1995); *3*, Amaya et al. (1991); *4*, Christian and Moon (1993); *5*, MacNicol et al. (1993); *6*, Bhushan et al. (1994); *7*, Cornell and Kimelman (1994); *8*, Graff et al. (1994); *9*, Holt et al. (1994); *10*, He et al. (1995); *11*, O'Reilly et al. (1995); *12*, Lamb et al. (1993); *13*, Ferreiro et al. (1994); *14*, Hemmati-Brivanlou et al. (1994); *15*, Hemmati-Brivanlou and Melton (1994); *16*, Witta et al. (1995); *17*, Blitz and Cho (1995); *18*, Lai et al. (1995); *19*, Taira et al. (1995); *20*, Ruiz i Altaba et al. (1993); *21*, Roelink et al. (1994); *22*, Lee et al. (1995); *23*, Hopwood and Gurdon (1990); *24*, Ludolph et al. (1994); *25*, Cary and Klymkowsky (1995); *26*, Ben Aziz-Aloya et al. (1993a); Seidman et al. (1994); *27*, Shapira et al. (1994); *28*, Lu et al. (1992), Valtorta et al. (1995); *29*, Schaeffer et al. (1994); *30*, Alder et al. (1995); *31*, Heasman et al. (1994); *32*, McCrea et al. (1993); *33*, Olson et al. (1991); *34*, Paul et al. (1995); *35*, Rashbass et al. (1992); *36*, Morasso et al. (1993); *37*, Rollins et al. (1993); *38*, Leibham et al. (1994); *39*, Seidman et al. (1995).

Stage	I	II	III	IV	V	VI
Size, uM	50-300	300-450	450-600	600-1000	1000-1200	1200
Morphology	Transluscent Oval	Opaque White, Oval	Even light brown, Round	Uneven light brown	Clearly defined Hemispheres	Equatorial belt
rRNA synthesis		+	+	+	+	+
mRNA synthesis	+	+	+	+	+	+
Mitochondrial production and organization	+	+	+	+	+	+
Vitellogenesis			+	+	+	

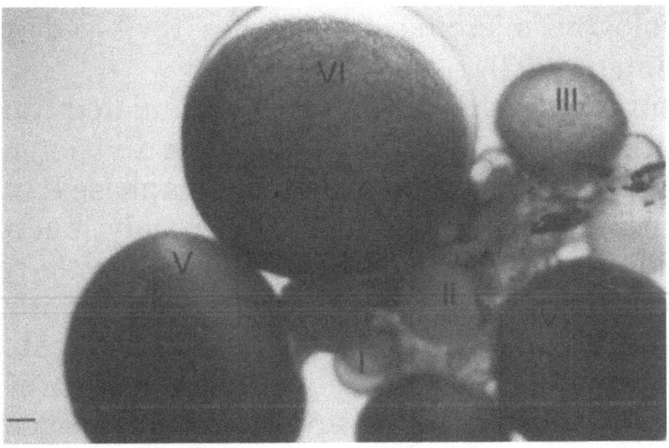

Fig. 1. *Xenopus* oogenesis. The stages of oocyte development (Dumont, 1972) are presented. Included are a description of the visible external markers and significant accompanying intracellular events. The exceptionally high rate of rRNA synthesis and accumulation occurring at stage 3 is accomplished through a transient 1500-fold amplification of the *Xenopus* rRNA genes (Hausen and Riebesell, 1991). Maternal stores of mRNA are established during stages 1–3, after which synthesis slows to maintain steady-state levels. A large portion (approx 90%) of maternal oocyte mRNA stores is translationally inactive. Mitochondria are amassed throughout oogenesis and organized into Balbiani bodies. Vitellogenesis marks the production and storage of yolk platelets from vitellogenin derived from the maternal blood supply. Stage 5 or 6 oocytes are most commonly used for microinjection experiments. Sizes of crosses in table indicate levels of activity. Size bar, 0.5 μm.

and polyA$^+$ messenger RNA [mRNA]), mitochondria, and yolk are accumulated for the needs of the early embryo. A salient feature of *Xenopus* oogenesis is the establishment of a distinct animal-vegetal polarity. Polarity in the oocyte is perceived externally as the hemispheric segregation of pigment. Internally, polarity is evidenced by the cytoskeletal organization (Palecek et al., 1985; Klymkowsky et al., 1987), in the unequal distribution of specific maternal mRNAs (Weeks and Melton, 1987; Ding and Lipshitz, 1993; Ku and Melton, 1993), and in the distribution of ion channels and receptors (Dascal, 1987; Oron et al., 1988). Oogenesis results in the formation of a mature, but unfertilizable oocyte arrested in prophase I of meiosis. Stage 5 and 6 oocytes are those most commonly employed for heterologous microinjection studies (reviewed by Soreq and Seidman, 1992).

Maturation is the progesterone-stimulated transition from prophase I through the first meiotic division to yield a mature fertilizable egg arrested in metaphase II of meiosis. Maturation is mediated by the autocatalytic activity of maturation-promoting factor (MPF), a cell cycle-regulating factor composed of $p34^{cdc2}$ and cyclin (Smith, 1989; Moreno and Nurse, 1990). In the laboratory, maturation may be stimulated in vivo by a single injection of gonadotropin or in vitro by the application of progesterone (Smith et al., 1991). A mature egg is clearly distinguished from an oocyte by the appearance of a prominent white spot on the animal pole representing the aftermath of germinal vesicle (nuclear) breakdown (GVBD).

Fertilization

Fertilization of several hundred eggs can be readily achieved in the laboratory either in vitro or through natural mating. Sperm entry, restricted to animal pole, triggers a rapid, transient, electrical "fast block to polyspermy" that is followed by a loss of sperm receptors from the membrane and the establishment of the "fertilization membrane," a final mechanical block to further sperm entry. Fertilization triggers the second meiotic division and the subsequent restoration of zygotic diploidy through fusion of the male and female pronuclei (beautifully illustrated in Hausen and Riebesell, 1991). Sperm entry marks the future ventral side of the embryo by initiating a series of events that break the radial symmetry of the egg and ultimately determine the dorsal/ventral embryonic axis (*see* Cortical Rotation section in this chapter).

Postfertilization

The most commonly used system for staging developing *Xenopus* embryos appears to be that established by Niewkoop and Faber (1967,1994) and is therefore the one used in the following description (Fig. 2). For detailed descriptions, diagrams, and histological illustrations of the morphogenetic transformations taking place during the early stages of *Xenopus* development (up to stage 27), the reader is again referred to the histological atlas of Hausen and Riebesell (1991) as well as Keller (1991) and Gilbert (1994).

Cortical Rotation

Approximately 1 h postfertilization, an internal, invisible-to-the-eye, microtubule-dependent rotation of the cortical cytoplasm takes place. The cortical rotation plays a critical role in the establishment of dorsal/ventral polarity in the embryo, presumably by exposing some yet unidentified maternal factor(s) (Sive, 1993; Slack, 1994). Disruption of the cortical rotation (e.g., by UV irradiation) precludes formation of a dorsal/ventral axis and leads to "hyperventralized" embryos. The ability to "rescue" UV-irradiated embryos has become one of the standard tests for identifying potential molecular participants in normal axial patterning in *Xenopus* (Smith and Harland, 1992; Ku and Melton, 1993; Steinbesser et al., 1993; He et al., 1995; *see also* Lemaire et al., 1995).

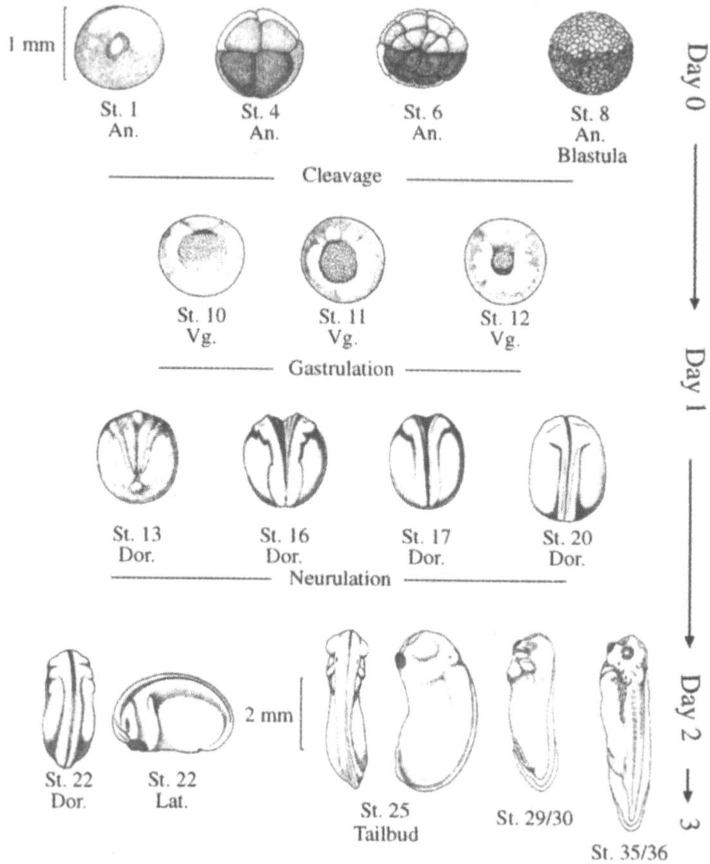

Fig. 2. Early *Xenopus* development. Significant stages marking the first 2–3 d of *Xenopus* development are presented. An, animal pole; Dor, dorsal; Lat, lateral; Vg, vegetal pole. Note that rate of development is highly temperature-dependent. Days indicated are approximate and based on incubation at 19–20°C. Adapted from Nieuwkoop and Faber (1994) with permission from Garland Publishing.

Cleavage (Stages 1–8 Leading to Blastula)

The first cleavage takes place approx 1.5 h postfertilization and is followed by 12 successive, synchronous cleavage cycles that occur at approx 1/2-h intervals. From the 13th cleavage, synchrony is lost concomitant to the onset of zygotic transcription. This milestone in early *Xenopus* development has been termed the "midblastula transition" (MBT; Newport and Kirschner, 1982a) and is likely regulated, at least in part, by a titratable factor (Newport and Kirschner, 1982b). Under normal experimental conditions, exogenous deoxyribonucleic acids (DNAs) appear subject to the same pre-MBT block to transcription as the zygotic DNA (Newport and Kirschner, 1982b; Krieg and Melton, 1985; Christian and Moon, 1993).

During these very early stages in *Xenopus* development, the first episodes in axial patterning and tissue differentiation take place. At this point, the animal hemisphere represents prospective ectoderm (i.e., the future nervous system) and the vegetal hemisphere, prospective endoderm (i.e., the future gut). Formation of the blastula coincides with initiation of a process called "mesoderm induction," in which equatorial ("marginal zone") cells are induced to form mesoderm in response to signals from the underlying vegetal cells (Fig. 3). In a widely accepted model in which ventral mesoderm is considered the ground or default state, two types of signals are initially active: a widespread general mesoderm-inducing signal and a localized dorsal mesoderm-inducing signal (reviewed by Kimelman et al., 1992; Slack, 1994) (Fig. 3). Evidence is accumulating, however, for the involvement of an additional, active, ventral mesoderm-promoting signal(s) that would serve to resist progressive dorsalization in the marginal zone and sharpen dorsal/ventral boundaries (Christian and Moon, 1993; Sive, 1993; Graff et al., 1994; He et al., 1995) (Fig. 3).

The general (and ventralizing) mesoderm-inducing signals promote differentiation of ventrolateral mesoderm, which will give rise to mesothelium, mesenchyme, and blood cells. The dorsal mesoderm-inducing signal induces axial mesoderm (notochord and segmented muscle) by promoting the formation of the Spemann organizer (Fig. 3). The organizer region serves to mediate diversification of dorso-ventral mesoderm and further plays a decisive role in neural-induction and antero-posterior axial patterning

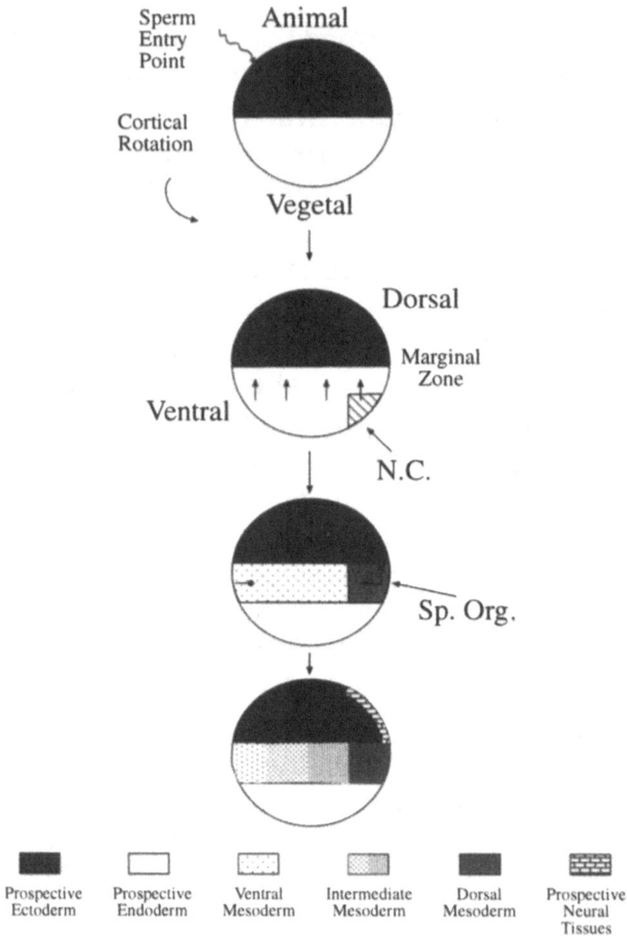

Fig. 3. Mesoderm induction. Schematic representation of the "three plus" signal model for mesoderm induction (Sive, 1993). Following fertilization, an approx 30° rotation of the cortical cytoplasm takes place, specifying the future dorsal/ventral axis in the embryo. Inducing factors (arrows) from the vegetal pole cells act upon marginal zone cells to induce formation of mesoderm. A specialized dorsovegetal region called the Nieuwkoop Center (NC) emits a dorsalizing signal, which is active from the 32-cell stage to around the midblastula stage and promotes formation of a second dorsalizing center known as the Spemann organizer (Sp. Org.). The organizer region becomes the principal dorsalizing influence from the mid–late blastula stage and promotes formation of notochord, axial muscle tissue, and the neural tube. Ventral mesoderm-promoting factors (round headed arrow) resist dorsalization and help define dorsal/ventral boundaries.

of the neural tube (Spemann and Mangold, 1924; Slack and Tannahill, 1992; Sive, 1993; Slack, 1994). Prominent players in mesoderm induction in *Xenopus* appear to include members of the transforming growth factor β (TGF-β) (Vg-1 and activin) and fibroblast growth factor (FGF) families of growth factors together with members of the Wnt family of developmentally expressed glycoproteins (Sive, 1993; Slack, 1994). The subset of mesoderm inducers that emanates from the dorsal-vegetal domain often referred to as the Nieuwkoop Center (Fig. 3) is considered likely responsible for generation of the organizer and may include Vg-1 and Wnt-11 (Sive, 1993; Slack, 1994). Whereas the Nieuwkoop Center is fully active from the 32-cell stage, the Spemann organizer displays significant activity around the MBT (Sive, 1993).

The response of cells in the marginal zone to mesoderm-inducing signals is evidenced, at least in part, by the localized expression of various putative transcription factors such as brachyury and snail (general mesoderm); goosecoid, forkhead, Xlim-1, noggin, Xnot, and pintallavis (organizer region); and wnt-8 and xpo (ventral mesoderm) (reviewed by Slack, 1994). These factors may further promote differentiation of specific cell types via dose-dependent or synergistic mechanisms. For example, ectopic overexpression of brachyury induces expression of the ventral marker α smooth muscle actin in animal cap explants (*see* Microinjection Strategies section in this chapter) up to 1 ng injected RNA (O'Reilly et al., 1995). However, higher amounts of microinjected RNA (up to 2.4 ng) induced expression of the dorsal mesodermal marker cardiac muscle actin. When brachyury was coexpressed with noggin (Cunliffe and Smith, 1994) or pintallavis (O'Reilly et al., 1995), notochord was also induced.

Gastrulation (Stages 8–13)

Gastrulation marks the transformation of the blastula, by a complex series of morphogenic transformations and cellular movements, into an embryo with clear antero-posterior polarity in which the three principal germ layers (ectoderm, endoderm, and mesoderm) and presumptive gut cavity (archenteron) have been established (Fig. 4). Gastrulation is initiated at the dorsal blastopore lip in the organizer region, proceeds laterally to form a ring of involuting cells, and concludes with the formation and closure of the yolk plug at the vegetal pole (*see* Keller, 1991 for a comprehensive and authoritative description of gastrulation in *Xenopus*). During gastrulation, prospective ectodermal cells from the animal pole spread vegetally through epibolic movements to encompass most of the external surface of the embryo. The vegetal edge of this ectodermal tissue is known as the noninvoluting marginal zone (NIMZ). The NIMZ extends downward to seal the yolk plug and later contributes to the formation of posterior neural structures. Following the lead of the bottle cells, superficially located prospective endodermal cells from the involuting marginal zone (IMZ) migrate to the inside of the embryo to form the roof of the archenteron. The deep cells of the IMZ comprise the prospective mesoderm and are internalized next: First come the migrating mesodermal cells, which move inward and upward to form the leading edge of the mesodermal mantle. These cells ultimately give rise to the head mesoderm. Next come the cells representing the future axial mesoderm. Upon involution, these cells undergo a process of convergence (constriction) and extension (elongation). Converging and extending mesoderm will ultimately form the notochord and somitic muscle, and participates, through inductive signaling, in the antero-posterior patterning of the neural tube (*see* the next section). The remaining vegetal cells represent prospective endoderm and form the future floor of the archenteron.

Neurulation (Stages 12–20)

These stages mark the formation of the neural plate and its subsequent folding and closure to form the neural

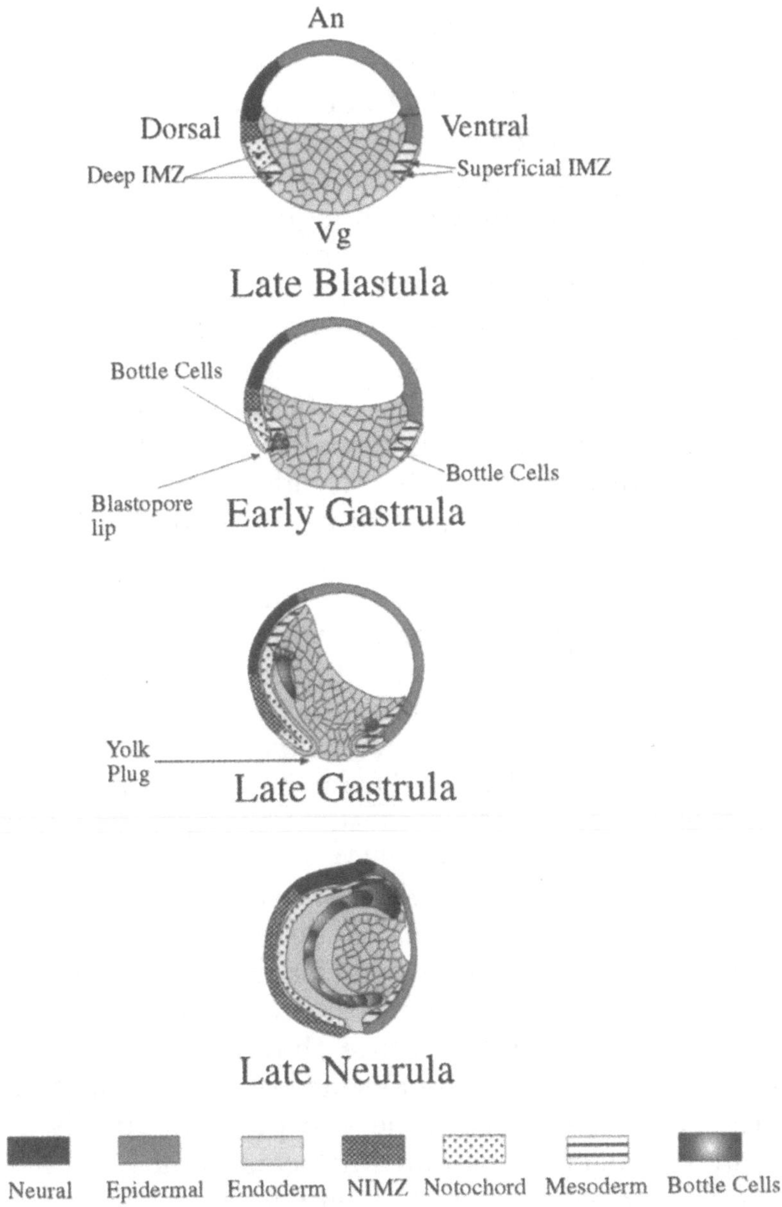

Fig. 4. *Xenopus* gastrulation. Morphogenetic movements occurring during gastrulation and neurulation are shown in simplified sagittal sections of late blastula, early and late gastrula, and late neurula-stage embryos. Prospective fates of tissues are indicated by shading as defined at the bottom of the figure. Dorsal is to the left and ventral is to the right. Adapted from Keller (1991).

tube. Antero-posterior polarity in the embryo is established during gastrulation, prior to the visible external appearance of the neural folds (Slack and Tannahill, 1992). Traditionally, gastrulation movements, bringing involuting mesoderm into direct physical contact with competent dorsal ectodermal cells, were thought to provide a vertical induction signal that was sufficient to induce and pattern neural tissue (Mangold, 1933; Eyal-Giladi, 1954) (Fig. 5). However, the observation that dorsal ectoderm can express neural markers in the absence of underlying axial mesoderm (Keller and Danilchik, 1988; Ruiz i Altaba, 1992) indicated that planar signals emanating from the organizer region and diffusing laterally through the dorsal ectoderm could play a significant role in neural induction as well (reviewed by Ruiz i Altaba, 1994). In classical treatments of neural induction, epidermis is considered the default destiny of prospective ectoderm in the absence of an "instructive" message from the dorsal mesoderm to "form neural tissue." However, a modified version of this model has been put forth in which neural induction results from a "permissive" signal derived from the mesoderm and consisting of inhibition of a neuralization repressing factor (Hemmati-Brivanlou et al., 1994).

Two molecules that are expressed in the organizer region and considered to participate in early neural induction events are noggin (Smith and Harland, 1992; Lamb et al., 1993) and follistatin (Hemmati-Brivanlou et al., 1994). Noggin is a secreted polypeptide that displays "direct" neuralizing activity in the absence of mesoderm and is expressed at the correct time and place to qualify as a potential endogenous neuralizing factor (Lamb et al., 1993). Follistatin, an antagonist of activin, was similarly shown to promote expression of neural markers in explants where no mesodermal differentiation could be demonstrated (Hemmati-Brivanlou et al., 1994). These examples of "direct" neuralizing activity support the contention that mesoderm induction is not an absolute prerequisite to neu-

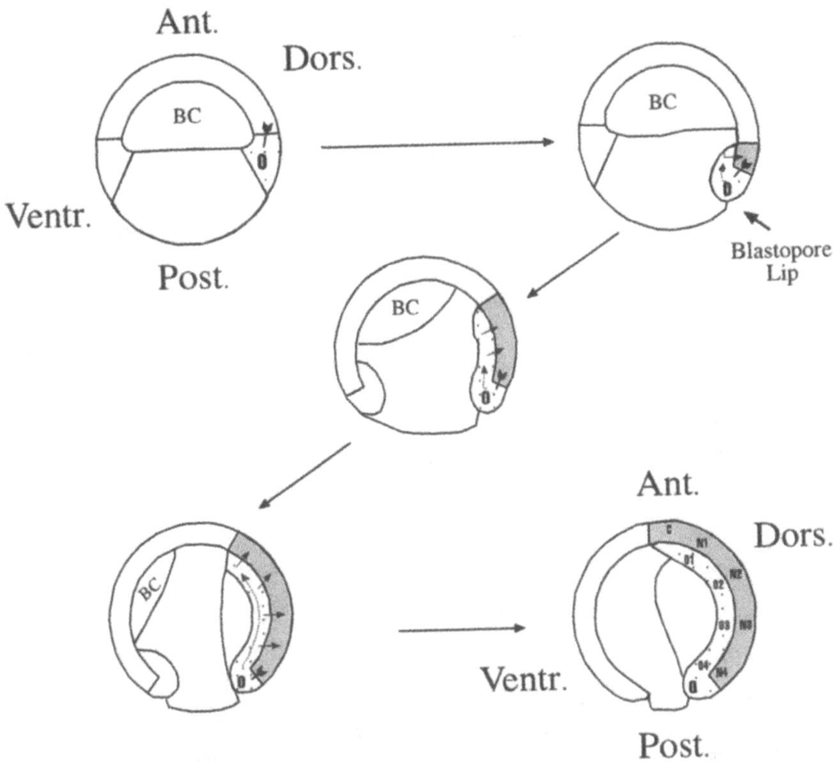

Fig. 5. Neural induction in *Xenopus*. Schematic representation of neural induction is presented beginning at late blastula stage. The organizer region (O) emits a planar signal (inverted arrowhead), which diffuses laterally through prospective neural ectoderm (gray) to promote formation of neural tissue. As gastrulation proceeds, the involuting mesoderm (dotted) is brought to lie underneath the prospective ectoderm and promotes neural and antero-posterior differentiation through direct physical contact (vertical signaling). N1–N4 and O1–O4 indicate correspondence in axial position between involuting mesoderm (O) and the developing neural tube (N). Adapted from Slack and Tannahill (1992).

ral differentiation (Hemmati-Brivanlou and Melton, 1994). XlPOU-2, a member of the class III POU domain family, also exhibits neuralizing activity and appears to function downstream of noggin (Witta et al., 1995). XASH-3, a

Xenopus homolog of the *Drosophila* Achaete Scute genes, encodes a putative basic helix-loop-helix (bHLH) DNA-binding protein that displays neuralizing activity and appears to function subsequent to induction of ectodermal tissues by noggin or related factors (Ferreiro et al., 1994). When overexpressed in microinjected *Xenopus* embryos together with the "promiscuous" heterodimer-forming bHLH transcription factor XE12, XASH-3 promoted transient expression of the neural-specific markers neuronal cell adhesion molecule (N-CAM), neurofilament, and type III β-tubulin V (Ferreiro et al., 1994).

Xotx2, a *Xenopus* homolog of the *Drosophila* orthodenticle gene is induced by the organizer-specific protein goosecoid and is progressively expressed in anterior neuroectoderm in register with the advancing mesoderm during gastrulation (Blitz and Cho, 1995). Xotx2 expression therefore corresponds well with the models of vertical induction. The Hox genes, thought to play a role in specifying antero-posterior information in diverse species such as *Drosophila* and mammals, appear to play a role in patterning the central nervous system (CNS) during *Xenopus* development as well (Godsave et al., 1994). The genes comprising the *Xenopus* HoxB cluster are strongly expressed from late gastrulation through the end of neurulation (Godsave et al., 1994). These genes display, at least in part, overlapping domains of expression, with 3' genes expressed more anteriorly and 5' genes more posteriorly within the developing nervous system (Godsave et al., 1994; Slack and Tannahill, 1994). An assortment of characterized molecular markers for specific positions along the incipient antero-posterior axis includes XAG-1, opsin, En-2, Tanabin, Krox-20, and XlHbox6 (Lamb et al., 1993; Hemmati-Brivanlou and Melton, 1994; Hemmati-Brivanlou et al., 1994).

Myogenesis (Stage 10+ Onwards)

Skeletal muscle is derived primarily from dorso-lateral mesoderm. Thus, skeletal muscle actin ("cardiac actin") is considered one of the classic markers for dorsal mesoderm induction (Lamb et al., 1993; Hemmati-Brivanlou et al., 1994; O'Reilly et al., 1995). As is true for other vertebrates and mammals, myogenesis in *Xenopus* is probably regulated by the MyoD family of myogenic regulatory factors (reviewed by Sassoon, 1993; *see also* Gurdon et al., 1989; Hopwood and Gurdon, 1990; Hopwood et al., 1991; Ludolph et al., 1994). Nonetheless, the spatio-temporal expression of MyoD in *Xenopus* differs significantly from that observed in mice and birds (Harvey, 1990; Frank and Harland, 1991; Sassoon, 1993). Firm commitment to skeletal muscle lineage probably occurs during or just after gastrulation, and almost all skeletal muscle derives from the somitic myotomes (Sasoon, 1993).

Somitogenesis (Stage 17 to Tadpole)

Somitogenesis is the process of segmentation of paraxial mesoderm, which gives rise to the myotomal muscle tissue. The first somite forms at stage 17 at the anterior trunk region of the embryo, and segmentation proceeds in a rostral-caudal direction at a rate of about 1 somite/40 min. Somitogenesis is characterized by the sequential detachment of blocks of cells (about 10 cells across) from the mesodermal ridge followed by a coordinated 90° rotation (Fig. 6). The length of a somite and thus, the length of a myotome and its muscle cells will therefore be 1 cell long and several cells thick.

The primary site of muscle innervation, which begins around stage 21, is at the ends of the muscle cells, which together form the intersomitic junction (Cary and Klymkowski, 1995). The high incidence of neuromuscular junctions (NMJs) at the intersomitic junctions gives rise to the characteristic banding observed in myotomal preparations histochemically stained for acetylcholinesterase (AChE) or acetylcholine receptor (AChR) (Figs. 7 and 8). By stage 24, essentially 100% of muscle cells demonstrate acetylcholine (ACh) sensitivity, spontaneous electrical activity, and contractibility (Kullberg et al.,

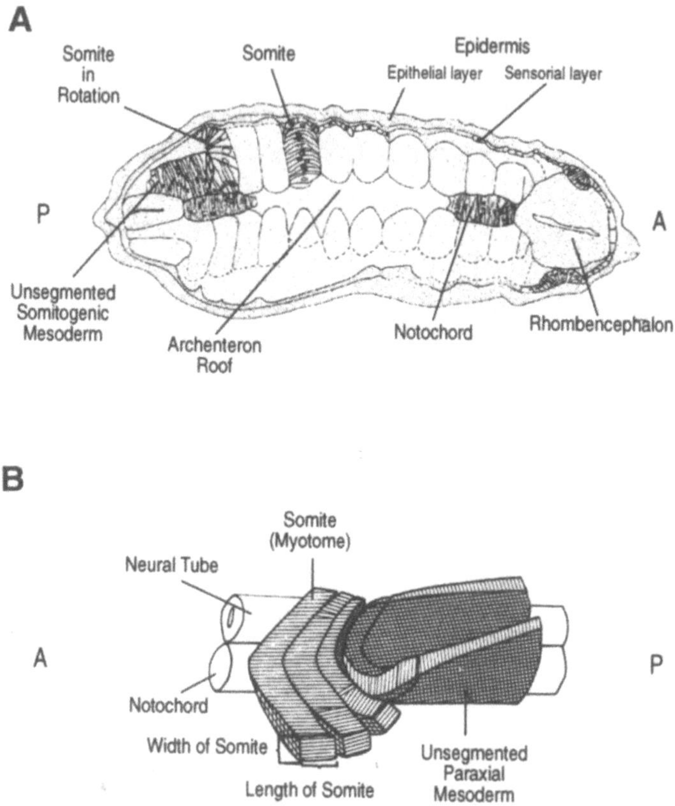

Fig. 6. *Xenopus* somitogenesis. **(A)** Sketch represents a horizontal section of a stage-22 embryo. Figure depicts the pattern of somites established at the anterior (A) end of the embryo and a somite in rotation at its posterior (P). Reprinted from Hausen and Riebesell (1991) with permission from Springer Verlag. **(B)** A three-dimensional reconstruction of somite formation in *Xenopus* is presented. The axial and paraxial structures are viewed from a ventro-lateral position on the left-hand side. Blocks of paraxial mesodermal cells are shown in the process of reorientation that marks somite formation. Adapted from Hamilton (1969).

1977). Furthermore, at these early stages, myotomal cells are electrically coupled—primarily within, but also between, individual somites (Kullberg et al., 1977; Cohen, 1980).

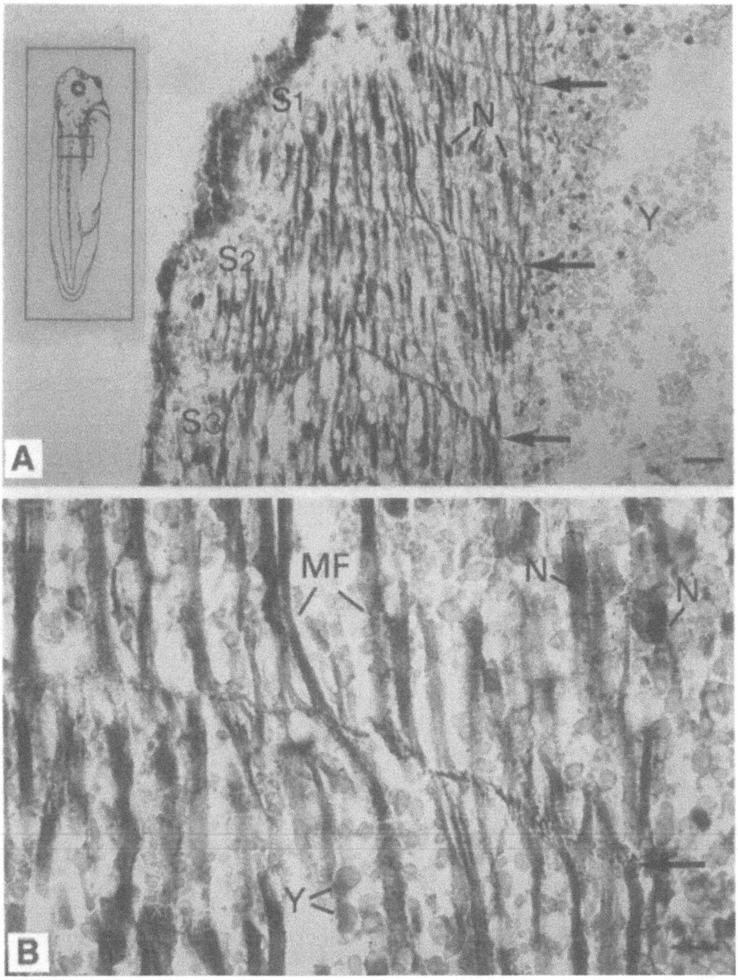

Fig. 7. Intersomitic AChE. Light microscope view (sagittal section) of the myotomal intersomitic junction of a stage 35/36 *Xenopus* embryo (inset) stained in whole mount for catalytically active AChE as described in Experimental Methodologies in Chapter 2, sectioned, and counterstained. Arrows indicate intersomitic junction; note the dark brown banding at the junction, which reveals accumulations of active enzyme. Motoneurons infiltrate the myotomes through the junctions, enriched in NMJs. MF, myofibers; N, nuclei; S, somite; Y, yolk platelets. Staining is hematoxylin-eosin-phosphomolybdic acid/light green stain. Histology courtesy of Asher Meshorer, Weizmann Institute of Science, Rehovot, Israel. Photographs were taken using a Zeiss Axioplan microscope fitted with the MC-100 camera (magnification **[A]** ×512 and **[B]** ×1600).

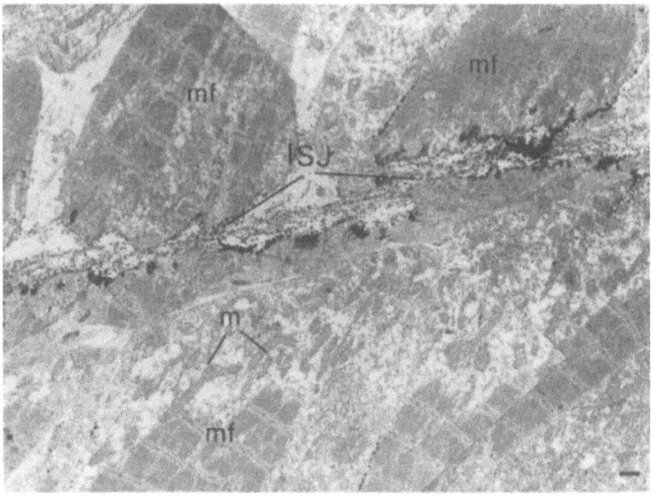

Fig. 8. Electron microscope view of the intersomitic junction. Longitudinal section of myotomal muscle from a 2-d-old *Xenopus* embryo (stage 25) stained for catalytically active AChE as described in Experimental Methodologies in Chapter 2. Sites of AChE accumulation appear as black crystals; note that parallel arrays of muscle cells (m) with their already well-developed myofibrils (mf) abut at the intersomitic junction (ISJ), which is characterized by the deposition of fibrous extracellular material and high accumulations of AChE. Size bar, 1 µm.

Hatching (Stage 25)

Following a period of spontaneous twitching, the tailbud embryo, approx 4 mm long, breaks out of the fertilization membrane. Hatching is probably achieved by a combination of mechanical and enzymatic efforts (Deuchar, 1966,1975). At this stage, the embryo responds with reflexive twitching to prodding with a sharp instrument.

Neuromuscular Junction Formation in Developing Xenopus *Embryos*

In the developing *Xenopus laevis* embryo, nerve and muscle differentiation, primitive neuromuscular contacts, and spontaneous synaptic activity are observed within 1 d postfertilization (Kullberg et al., 1977). During the ensuing 24 h, ultrastructural specializations characterizing synaptic differentiation are observed, followed by the acquisition of spontaneous motor activity and hatching (Cohen, 1980). Fervent embryonic development and ultrastructural maturation of the neuromuscular system continue, giving rise to a free swimming tadpole within 4–5 d. The rapid development of the neuromuscular system in *Xenopus* thus makes it an excellent in vivo model for transgenic studies of vertebrate synaptogenesis.

In vivo, the acquisition of ACh sensitivity precedes nerve–muscle contact and synaptic activity by at least 1 h (Blackshaw and Warner, 1976); spontaneous electrical activity can be detected almost immediately upon nerve outgrowth into the muscle tissue (approximately stage 19; Kullberg et al., 1977). The earliest nerve–muscle contacts take place between stages 21 and 24 (Kullberg et al., 1977) (Figs. 9 and 10). These contacts, which represent primitive synapses, lack any of the morphological specializations of the mature NMJ. Significant increases in the number of nerve–muscle contacts and the extent of cytological and morphological differentiation are observed between stages 25 and 36. The first postsynaptic specializations reminiscent of NMJ maturation are evident around stage 27 and consist of thickened elevations of the postsynaptic membrane. By this time, >95% of the membrane associated α-bungarotoxin (α-BTX) binding sites are localized to the ends of the muscle cell (Cohen, 1980). NMJ development continues with the cytological maturation of the nerve terminal and the synaptic cleft with its associated basal lamina. The persistent decline in the time-course of spontaneous and evoked postsynaptic potentials, which accompanies neuromuscular development (Kullberg et al., 1980), probably reflects a developmentally regulated increase in AChE (Gindi and Knowland, 1979). By stages 48–52, NMJs appear to reflect all the properties characterizing a mature, albeit embryonic, junction.

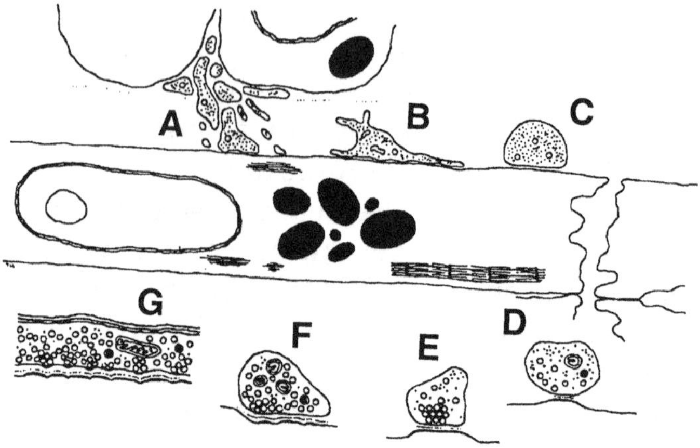

Fig. 9. *Xenopus* synaptogenesis. Schematic diagram depicting the significant morphological events in NMJ formation in *Xenopus.* **(A)** Motor nerve processes exit the spinal cord, penetrate into the myotomes at the intersomitic junctions, and establish initial primitive contacts with myofiber. **(B)** Processes resembling growth cones and bearing filopodia are situated between or on the surface of the myofibers. **(C)** Rounded processes devoid of filopodia are distanced from the muscle by a regular space. These processes contain high levels of glycogen and, occasionally, a synaptic vesicle. **(D)** Postsynaptic membrane appears elevated at the site of contact and acquires a thickened appearance. Dense material appears in the incipient synaptic cleft. Nerve terminals display increased numbers of vesicles and decreased levels of glycogen. **(E)** Clustering of vesicles at the presynaptic membrane. **(F)** Ridges appear on the postsynaptic membrane. Widening of the synaptic cleft. **(G)** Widening of the nerve–muscle contact surface. Appearance of presynaptic active zones. Maturation of the junctional folds (stage 50) (Kullberg et al., 1977).

Cultures of myotomal muscle cells are readily prepared from stage 20–22 *Xenopus* embryos (Evers et al., 1989; Peng et al., 1991; Alder et al., 1992; Lu et al., 1992). In culture, noninnervated myotomal cells display colocalized patches of AChE and AChR, which occupy about 3% of the cell surface (Moody-Corbett and Cohen, 1981). In addition, noninnervated cells exhibit localized ultrastructural specializations resembling those of the postsynaptic membrane. When myotomal muscle cells are cocultured

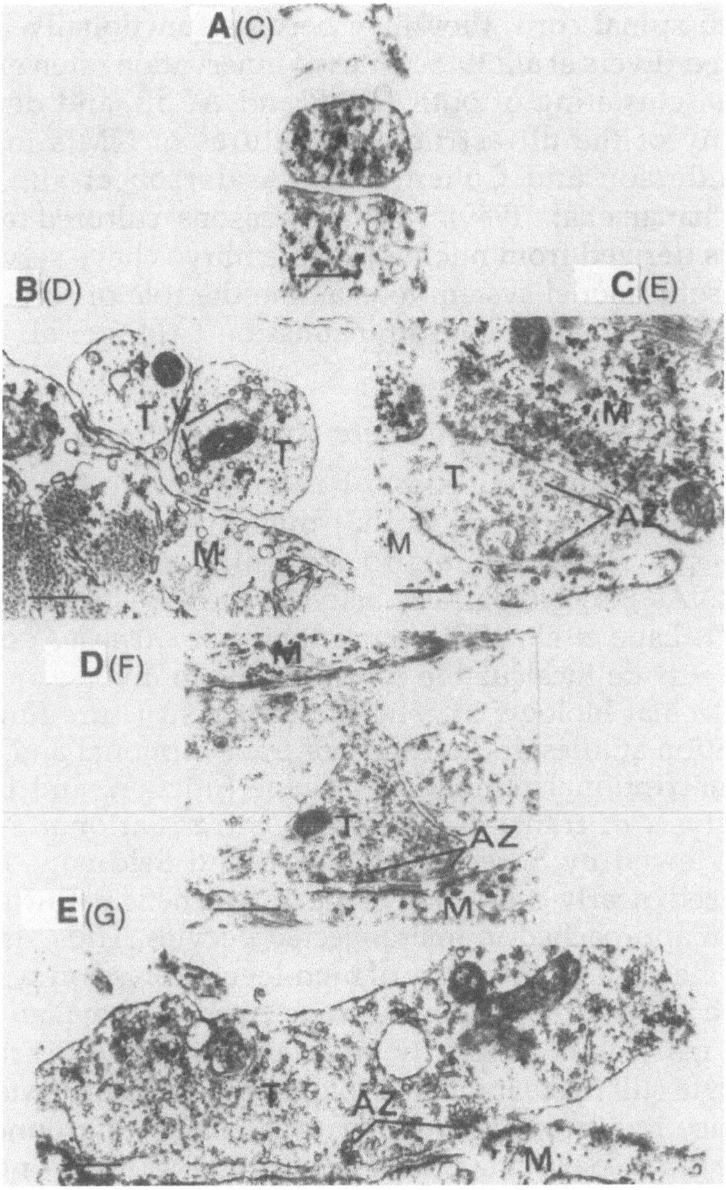

Fig. 10. Developing *Xenopus* NMJs. Electron micrographs of representative NMJs found in myotomes of 2–3-d-old *Xenopus* embryos. Letters in parentheses reflect approximate correspondence to stages described in Fig. 9. AZ, active zones; M, muscle cell; T, nerve terminal; V, synaptic vesicles. Size bar, 0.5 μm.

with spinal cord, they may become functionally innervated (Evers et al., 1989). Sites of innervation often exhibit local clustering of both AChR and AChE, and develop many of the ultrastructural features of NMJs in vivo (Anderson and Cohen, 1977; Anderson et al., 1977; Buchanan et al., 1989). For these reasons, cultured muscle cells derived from microinjected embryos have served as a useful model system to examine the role of some proteins on synaptic neurotransmission (Alder et al., 1992; Lu et al., 1992).

Xenopus *Oocyte Microinjection*

A quarter of a century has passed since the revolutionary demonstration that microinjected oocytes of *Xenopus laevis* are able to faithfully translate foreign mRNAs into biologically active proteins (Gurdon et al., 1971; Lane et al., 1971). Since that time, *Xenopus* oocytes have made incalculable contributions to diverse areas of molecular biology: in gene cloning, in structure-function relation studies, in the study of transcriptional and posttranscriptional regulation of gene function, and in the analysis of translational and posttranslational events (reviewed by Soreq, 1985; Soreq and Seidman, 1992). Indeed, nearly every step in the biosynthetic pathway has been approached in microinjected oocytes. Thus, despite the increasing popularity of high-level expression systems such as bacteria, yeast, and transfected mammalian cells, the marvelous simplicity and versatility of the *Xenopus* oocyte still make it the heterologous expression system of choice in a growing number of laboratories around the world. In particular, investigators working in neurosciences are intensively exploiting *Xenopus* oocyte microinjection in the study of receptors, ion channels, pumps, and transporters.

Oocyte expression protocols include the microinjection of total or size-fractionated polyA^+ RNA, in vitro-transcribed RNA, and plasmid DNA (Soreq and Seidman,

1992). In comparison to in vitro translation systems such as reticulocyte lysates or wheat germ extracts, oocytes perform more rounds of translation/molecule of mRNA and perform, at least partially, posttranslational modifications such as glycosylation, phosphorylation, and the removal of signal or other propeptide sequences (Soreq, 1985). As an intact cell, oocytes offer the opportunity to examine compartmentalized cellular processes such as the nuclear or cortical localization of RNA, subcellular protein transport, and secretion. For the molecular neurobiologist, the expression of biologically active receptors, channels, and ion pumps offers an excellent opportunity to study the assembly and membrane insertion, as well as ligand-binding and conductance properties, of this important family of molecules (Snutch, 1988; Soreq and Seidman, 1992; Patrick et al., 1993). The outstanding feature of heterologous channels and receptors expressed in oocytes is that they often display the pharmacological and electrophysiological characteristics of the native molecules in their natural milieus. This latter quality permits the accurate discernment and characterization of receptor subtypes (Barnard, 1988; Lester, 1988) and facilitates structure-function studies through site-directed mutagenesis and oocyte expression (*see* Table 2). The technical aspects of the use of microinjected *Xenopus* oocytes for expression of ion channels have been comprehensively reviewed (Rudy and Iverson, 1992).

Cholinergic proteins have been extensively studied using microinjected *Xenopus* oocytes and provide an excellent example for the variety of experimental strategies that can be used to clone and characterize important nervous system proteins (Table 2). Indeed, the *Torpedo* nicotinic AChR (nAChR) was the first ligand-gated ion channel to be functionally expressed in oocytes (Mishina et al., 1984). By now, almost all the primary elements comprising the diverse cholinergic synapses of the central and peripheral nervous systems have been expressed in

Table 2
Xenopus Oocyte Microinjection Strategies

Experimental strategy	Principal conclusions	Ref.
Expression of functional receptor or enzyme		
PolyA$^+$ mRNA from *Torpedo* electric organ or rat brain	First heterologous expression of catalytically active AChE	Soreq et al. (1982) *Proc. Natl. Acad. Sci. USA* **79,** 830
PolyA$^+$ RNA from *Torpedo* electric organ	First heterologous expression of active ChAT	Gundersen et al. (1985) *Proc. Natl. Acad. Sci. USA* **82,** 608
Size-fractionated polyA$^+$ mRNA from primary brain tumors	Evidence for human AChE mRNA polymorphism	Soreq et al. (1984) *EMBO J.* **3,** 1371
In vitro-transcribed mRNA encoding *Torpedo* nAChR α-, β-, γ-, δ-subunit mRNAs	α subunit sufficient for α-BTX binding; all four subunits sufficient and necessary for electrophysiological response to ACh	Mishina et al. (1984) *Nature* **307,** 604
In vitro-transcribed mRNA encoding muscarinic AChR	Identification of at least five pharmacologically distinct muscarinic subtypes	Fakuda et al. (1987) *Nature* **327,** 623
In vitro-transcribed mRNA encoding mammalian ChAT	Verified ChAT coding sequence	McCaman et al. (1988) *Mol. Brain Res.* **3,** 107
In vitro-transcribed mRNA encoding human AChR	Verified ACHE gene coding sequence; native ligand-binding properties; oligomeric assembly	Soreq et al. (1990) *Proc. Natl. Acad. Sci. USA* **87,** 9688
In vitro-transcribed mRNA encoding rat choline transporter	Verified CHOT1 gene coding sequence	Mayser et al. (1992) *FEBS* **305,** 31

α7 neuronal nAChR cDNA under SV40 promoter	α7 can produce a functional homo-oligomer with high-calcium permeability	Couturier et al. (1990) *Neuron* **5,** 847
AChE-encoding DNA behind native human promoter or CMV enhancer-promoter	CMV promoter active in *Xenopus;* human ACHE gene promoter expressible in *Xenopus;* correct splicing of intron I1	Ben Aziz-Aloya et al. (1993a) *Proc. Natl. Acad. Sci. USA* **90,** 2471
Subunit mixing experiments		
Substitution of Mouse nAChR δ subunit for *Torpedo* δ subunit	Demonstrated functional cross-species homology	White et al. (1985) *Proc. Natl. Acad. Sci. USA* **82,** 4852
Coinjection of calf nAChR δ with *Torpedo* α, β, γ subunits	δ subunit confers species-specific desensitization characteristics	Sakmann et al. (1985) *Nature* **18,** 538
Substitution of calf-specific nAChR ε subunit with adult-specific δ subunit	δ/ε subunits dictate differences between fetal and adult channel properties	Mishina et al. (1986) *Nature* **321,** 406
Substitution of mouse β1 muscle nAChR for β2 neuronal nAChR	Demonstrated interchange ability of muscle/neuronal β subunits	Deneris et al. (1988) *Neuron* **1,** 45
Coinjection of *Torpedo* α-, β-, or γ-subunit mRNA with cat polyA$^+$ mRNA	δ subunit confers species specific desensitization properties	Sumikawa and Miledi (1989) *Proc. Natl. Acad. Sci. USA* **86,** 367
Compared gating modes of α_2, β, γ, δ with α_2, β, γ, ε rodent nAChR	Subunit-specific stabilization of intrinsic fluctuating gating modes may account for developmental changes in receptor kinetics	Naranjo and Brehm (1993) *Science* **260,** 1811

(continued)

Table 2 *(continued)*

Experimental strategy	Principal conclusions	Ref.
Pairwise coinjection of various α and β subunit neuronal nAChR mRNAs	Different combinations of subunits give rise to receptors with distinct pharmacological and electrophysiological properties	Reviewed by Patrick et al. (1993) *Prog. Br. Res.* **98,** 115
Site-directed mutagenesis		
Site-directed mutagenesis of Leu247 in α7 neuronal AChR	Leu247 is a channel-facing residue in MII domain; mutations reduce desensitization, increase affinity for ACh and abolish current rectification; mutations may render a desensitized state conductive	Revah et al. (1991) *Science* **353,** 846; Bertrand et al. (1992) *Proc. Natl. Acad. Sci. USA* **89,** 1261
Site-directed mutagenesis of mouse nAChR δ subunit and coinjection with remaining subunits	δ-Asp180 and Glu189 important in ACh-binding of receptor	Czajkowski et al. (1993) *Proc. Natl. Acad. Sci. USA* **90,** 6285
Site-directed and deletion mutagenesis of m3 muscarinic receptor	Charged residues involved in coupling receptor to second messenger pathway	Kunkel and Perlta (1993) *EMBO J.* **12,** 3809
In vitro-transcribed mRNA encoding allelic and site-directed variants of human BuChE + inhibition profiles with commonly used drugs and poisons	Allelic point mutations alter sensitivity to inhibitors; predicted genetic variability to anti-ChE drug therapies	Loewenstein-Lichtenstein et al. (1995) *Nature Genet.* **10,** 1082

Expression of chimeric molecules		
Construction of chimeric γ–δ or α–δ mouse muscle nAChR subunits and coinjection with α, β, γ subunits	"Delta" character of δ subunit distributed over the entire polypeptide	Mixter-Mayne et al. (1987) *Mol. Brain Res.* **2,** 191
Coinjection of mammalian neuronal nAChR $\beta2/\beta4$ chimera with $\alpha4$ subunit mRNA	80 amino acid N-terminal domain confers neuronal bungarotoxin sensitivity and agonist selectivity	Wheeler et al. (1993) *FEBS* **322,** 139
Replacement of 76 amino acid peptide in human BuChE with corresponding peptide from AChE	Peptide domain comprising rim of active site gorge (amino acid 58 to 133) confers ligand-binding specificity	Loewenstein et al. (1993) *J. Mol. Biol.* **234,** 289
Coexpression of auxiliary factors		
Coinjection of in vitro-transcribed mRNA and brain or muscle polyA^+ mRNA	Tissue-specific assembly of human BuChE dependent on tissue-specific auxiliary factors	Soreq et al. (1989) *J. Biol. Chem.* **264,** 10,608
Coinjection of mouse nAChR mRNAs with mRNA encoding 43-kDa protein followed by immunofluorescence microscopy	43-kDa protein clusters nAChR	Froehner et al. (1990) *Neuron* **5,** 403

(continued)

Table 2 *(continued)*

Experimental strategy	Principal conclusions	Ref.
Site-directed mutagenesis of 43-kDa protein and coinjection with mouse nAChR receptor DNAs	C-terminal zinc finger domain of 43-kDa protein important for AChR clustering	Scotland et al. (1993) *J. Cell Biol.* **123,** 719
Tunicamycin to study glycosylation		
Coinjection of *Torpedo* nAChR subunit mRNAs and tunicamycin treatment of oocytes	Glycosylation required for correct plasma membrane insertion but not for oligomeric assembly	Sumikawa and Miledi (1989) *Mol. Br. Res.* **5,** 183
In vitro-transcribed-BuChE mRNA followed by immunohistochemistry and electron microscopy	Polarized extracellular deposition dependent on glycosylation	Dreyfus et al. (1989) *Cell. Mol. Neurobiol.* **9,** 323
Other		
Reporter epitopes and MAbs used to examine membrane topology of human nAChR $\alpha1$ subunit	Identification of cytoplasmic and extracellular domains	Anand et al. (1993) *Biochem* **32,** 5975

oocytes, including neuronal nicotinic (reviewed by Patrick et al., 1993) and muscarinic AChRs (Fakuda et al., 1987), the ACh synthesizing enzyme choline acetyltransferase (ChAT) (reviewed by Wu and Hersh, 1994), a choline transporter (Mayser et al., 1992), and the ACh hydrolyzing enzymes AChE and butyrylcholinesterase (BuChE) (Table 2).

In conclusion, oocyte expression studies have promoted rapid advances in our understanding of the structures and inner workings of important nervous system proteins. However, as a single, highly specialized, and relatively quiescent cell, the oocyte does not offer the biological diversity necessary to study their tissue-specific processing and biological roles. To address these questions, a more complex, dynamic expression system is required.

Xenopus *Embryo Microinjection*

Overview

To study the role and regulation of specific gene products in embryonic development, microinjected *Xenopus* embryos have been employed. Microinjected *Xenopus* embryos offer several advantages over other heterologous and transgenic expression systems. In comparison to oocytes or transfected mammalian cells, the embryos offer a dynamic, multicellular, in vivo environment representing numerous cell types passing rapidly through various stages of differentiation and organogenesis. In contrast to transgenic mice, the *Xenopus* system offers a rapid, convenient, economical, small vertebrate model with which it is possible to examine a large number of cloned DNA sequences in a statistically significant number of individuals within a time-frame of days to weeks. In the current age of molecular biology and protein engineering, the enormous versatility of microinjected *Xenopus* embryos in gene function assays is to be particularly appreciated.

A special issue of *Methods in Cell Biology* has been devoted to *Xenopus* oocytes and embryos as an experimental system (Kay and Peng, 1991). A wonderful Internet site for *Xenopus* research is the *Xenopus* Molecular Marker Resource created and maintained by Peter Vize's laboratory at the University of Texas, Austin (URL http://vize222.zo.utexas.edu/). The database includes a methods and protocols index, tested DNA sequence primers for reverse transcriptase-polymerase chain reaction (RT-PCR), and a collection of images illustrating the expression patterns of a wide variety of *Xenopus* genes as detected with both antibody and nucleic acid probes.

Owing to their large size, resilience, and accessibility, *Xenopus* embryos have long been a favorite specimen among classical embryologists. Indeed, the amphibian model—derived in great part from studies in *Xenopus*—probably offers the most sophisticated picture available today of early events such as mesoderm induction and axial patterning in vertebrate development (Gurdon et al., 1989; Sive, 1993; Slack, 1994). Classic embryology relied principally on mechanical micromanipulations to address questions regarding commitment, induction, and the role of cell–cell interactions in the impartation of positional information (Spemann and Mangold, 1924; Nieuwkoop, 1973). Now, decades later, molecular biologists are exploiting gene cloning and microinjection strategies to identify the molecules and biochemical pathways mediating these events (reviewed by Sive, 1993; Slack, 1994) (Table 1). Microinjection protocols have been particularly useful in elucidating the molecular pathways invoked during myogenesis and neural induction in *Xenopus* (Table 3, pp. 32–35; *see* Microinjection Strategies section in this chapter). The fact that many of the cloned *Xenopus* genes that appear to be active in early embryonic development display significant sequence and functional homology to the corresponding genes in both *Drosophila* and mammals underscores the high degree of evolutionary conservation among the molecules controlling early developmental events. Some *Drosophila* genes have even proven to be biologically active in *Xenopus* microinjection assays (Rothbacher et al., 1995; Tannahill et al., 1995). Moreover, functional conservation provided the basis for the cloning of human and rat homologs of noggin based on *Xenopus* expression studies (Valenzuela et al., 1995).

Microinjection Strategies

General Considerations

A typical embryo microinjection experiment will include priming several female frogs with human chorionic gonadotropin (HCG) 1 d prior to the experiment to induce spawning followed by in vitro fertilization and microinjection of 10–20 nL ($\leq = 1$ ng) RNA or DNA encoding the protein of interest (*see* Appendix V for a detailed protocol). Coinjection of an mRNA encoding an easily detectable protein such as β-galactosidase (Smith and Harland, 1992; Ferreiro et al., 1994; Witta et al., 1995) or a fluorescently labeled dextran (Gimlich, 1991; Ruiz i Altaba and Jessell, 1992) may be employed to trace embryonic cells that receive the heterologous sequence. The injected embryos are allowed to develop from overnight to several days and are assessed for expression of heterologous gene products and developmental effects. In an exciting technical advance regarding lineage tracing, DNA encoding the naturally fluorescent green fluorescent protein (GFP) was subcloned into a vector carrying a *Drosophila* E(spl) gene (Tannahill et al., 1995). This allowed the examination of live embryos under a fluorescent microscope and afforded the investigators the opportunity to select embryos with the desired RNA distribution. The choice of RNA or DNA for injections can be expected to influence the timing and levels as well as spatial distribution of expressed protein (*see* Vize et al., 1991 for a thorough discussion) and, therefore, deserves consideration.

Table 3
Neural and Myogenic Pathways Studied by Ectopic Expression in *Xenopus* Embryos[a]

Molecule(s) overexpressed	Whole embryo effects	Animal cap effects	Conclusions	Ref.
Pintallavis	Enhanced posterior neural development	ND	Pintallavis is a neural inducer	Ruiz i Altaba and Jessell (1992) *Development* **116,** 81
Pintallavis	Induced ectopic expression of floor-plate-specific marker F-spondin	ND	Pintallavis is involved in floor plate differentiation	Ruiz i Altaba et al. (1993) *Proc. Natl. Acad. Sci. USA* **90,** 8268
Noggin	ND	Induced neural marker N-CAM, but not mesodermal marker cardiac actin	Noggin displays expression pattern and biological activity to be an endogenous neural inducer	Lamb et al. (1993) *Nature* **262,** 713
Dominant-negative activin receptor	Exaggerated neural tissue formed from injected animal pole cells incl. multiple eyes; ectopic neural tissue; hypertrophy of normal CNS Relocation of injected vegetal cells to dorso-anterior position Partial rescue of UV-ventralized embryos	Expression of general neural markers in isolated animal caps, incl. N-CAM and β tubulin; Induced expression of antero-posterior axis-specific markers incl. opsin, En-2, tanabin, Drox-20, Xlhbox-6	Activin is a natural inhibitor of neuralization	Hemmati-Brivanlou and Melton (1994) *Cell* **77,** 273
Follistatin	ND	Blocked activin-induced morphogenetic movements Induced neural markers	Follistatin is a natural antagonist of activin which derepresses neuralization signals	Hemmati-Brivanlou et al. (1994) *Cell* **77,** 283

XASH-3 + XE12	Enlargement of neural tube	Transient activation of neural markers N-CAM, neurofilament, and β tubulin	XASH-3 acts as a proneural gene	Ferreiro et al. (1994) *Development* **120,** 3649
Vertebrate hedgehog (vhh-1)	Induced ectopic expression of floor plate markers F-spondin and HNF-3B	ND	Vhh-1 may contribute floor-plate inducing activities of notochord	Roelink et al. (1994) *Cell* **76,** 761
Orthodenticle (Xotx2)	Induced ectopic cement gland Endogenous Xotx2 induced by goosecoid	Enhanced expression of cement gland-specific marker XAG1 Induced general neural marker N-CAM	Xotx2 is a target of goosecoid which functions in anterior-posterior axis specification	Blitz and Cho (1995) *Development* **121,** 993
XIPOU-2	Induced ectopic expression of β tubulin in neurula stage embryos	Induced expression of β tubulin and N-CAM but not mesodermal markers Explants displayed neuronal phenotype incl. process outgrowth Did not induce noggin or follistatin	XIPOU-2 is a proneural gene which functions downstream of both noggin and follistatin	Witta et al. (1995) *Development* **121,** 721
Brachyury + pintallavis	ND	Xbra alone induces mesodermal markers in a dose-dependent manner Xbra + pintallavis induces notochord and neural tissue in dose-dependent manner	Pintallavis synergistically cooperates with brachyury to pattern mesoderm	O'Reilly et al. (1995) *Development* **121,** 1351

(continued)

Table 3 *(continued)*

Molecule(s) overexpressed	Whole embryo effects	Animal cap effects	Conclusions	Ref.
XMyoD or XMyf-5	Developed normally	Transient enhancement of cardiac muscle actin gene expression Autoinduction of endogenous MyoD No muscle differentiation	XMyoD can activate muscle specific genes but additional factors are required for complete and stable myogenesis	Hopwood and Gurdon (1990) *Nature* **347,** 197 Hopwood et al. (1991) *Development* **111,** 551
XMyoD + E12	ND	10-fold enhancement of induced cardiac actin gene expression over that obtained with XMyoD alone Activated expression of 12/101 antigen associated with differentiated skeletal muscle	MyoD and E12 cooperate to form a DNA-binding complex which interacts specifically with the cardiac actin gene promoter to activate expression during myogenesis	Rashbass et al. (1992) *EMBO J.* **11,** 2981
XMyoD or XMyf-5	Enlargement of myotomes via recruitment of nonsomitic cell lineages Ectopic muscle in nervous tissue Neural and epidermal abnormalities	Enhanced expression of cardiac muscle actin	MyoD and Myf-5 serve redundant functions MyoD and Myf-5 initiate developmental program for primary myotomal myocytes	Ludolph et al. (1994) *Devel. Biol.* **166,** 18
Synaptophysin	Synaptophysin involved in late steps of neurotransmitter release	ND	Increased frequency of spontaneous synaptic currents	Alder et al. (1995) *J. Neurosci.* **15,** 511

			Increased amplitude of evoke synaptic currents Increase rate of tetanus-induced fatigue	
Truncated vimentin or desmin	Disruption of intermediate filaments and abnormal organization at intersomitic junctions Invaginations of sarcolemma characteristic of myocyte termini reduced or absent	ND	Intermediate filaments play an important role in organization and/or stabilization of myofibril-membrane attachment sites	Cary and Klymkowsky (1995) *Development* **121,** 1041
	Attachment of myofibrils to junctional membrane disrupted	ND		

[a]Abbreviation: ND, not determined.

Despite the convenience of DNA injections, microinjection of in vitro-transcribed mRNAs appears to be preferred in many laboratories—probably owing to the observation that distribution of the ultimate protein product is more widespread when RNA is used. In addition, injected RNA will give rise to protein sooner than DNA, permitting studies on very early developmental events. Factors to consider when planning DNA or RNA microinjections should include the expression or transcription vector, 3' and 5' flanking sequences (e.g., polyA tail), concentration, and injection site (reviewed by Vize et al., 1991). Especially significant is the observation that some sequences may induce global deformities when injected at high concentrations or into a certain hemisphere; in lower concentrations or in the opposite hemisphere, these sequences may generate experimentally significant effects such as duplication of an embryonic axis (Ruiz i Altaba and Melton, 1989; Thomsen et al., 1990).

The fate of injected DNA depends on whether linear or supercoiled DNA is employed, and may include extrachromosomal replication, concatamerization, association with chromosomal DNA, and variable stability (Bendig, 1981; Rusconi and Schaffner, 1981; Etkin et al., 1984; reviewed by Vize et al., 1991). DNA-mediated expression will probably be initiated only at the midblastula transition and tends to be transient and mosaic in nature. A variety of eukaryotic promoters such as hsp70 (Harland and Misher, 1988), the murine sarcoma virus long-terminal repeat (Giebelhaus et al., 1988), cytomegalovirus (CMV) (Roelink et al., 1994; Seidman et al., 1994), and SV40 (Shapira et al., 1994) have been successfully employed to drive transcription in embryos. However, correct temporal and/or tissue-specific regulation of transcription from injected DNAs has been reported only in a few instances, and only where promoters from *Xenopus* genes were employed. In those cases where correct spatio-temporal expression of heterologous DNAs is observed, embryo microinjections can further be exploited to dissect regulatory elements directing stage- or tissue-specific gene expression (*see* the next section). Thus, microinjected *Xenopus* embryos may serve in studies of both gene function and regulation.

Studying Gene Regulation

The utility of *Xenopus* embryo microinjections for analyses of gene regulation has been the subject of a comprehensive review (Sargent and Mathers, 1991). For example, Krieg and Melton (1985) observed gastrulation-specific expression of the gastrula-specific protein GS17 in microinjected embryos and subsequently identified an enhancer responsible for activating transcription at the MBT (Krieg and Melton, 1987). The control element(s) specifying epidermis-specific transcription of an embryonic keratin was traced, using microinjected embryos, to a 5' flanking region of the XK81A1 gene, using β-globin as a reporter gene (Jonas et al., 1989). A regulatory sequence specifying epidermal expression of the Distal-less-like-2 (Xdll-2) homeodomain protein was subsequently identified using a similar strategy (Morasso et al., 1993). Following the demonstration that a plasmid carrying the *Xenopus* cardiac actin gene is subject to muscle-specific transcriptional controls in microinjected embryos (Wilson et al., 1986), the system was again exploited to pinpoint the control element(s)—this time a CArG sequence in the promoter (Mohun et al., 1989). The TATA box present in the *Xenopus* MyoD (XMyoD) gene was shown, in microinjected embryos, to play a crucial role in mediating somite-specific transcription of this myogenic regulatory protein (Leibham et al., 1994). In a powerful application of embryo microinjection to study tissue-specific gene expression, coinjection of RNA encoding the DNA binding protein E12 with that encoding XMyoD induced 10-fold greater induction of the cardiac actin gene than that achieved with MyoD-RNA alone (Rashbass et al., 1992).

Studying Gene Function by Overexpression

Induction Assays in Animal Caps. Explanted blastula-stage animal-pole cells (animal caps) continue to develop in culture and differentiate into amorphous epidermal tissue. When combined with marginal zone cells from the vegetal hemisphere or externally applied growth factors, explanted animal caps are induced to form mesodermal tissues, which display characteristic morphogenetic behaviors and express specific molecular markers. This feature allows for the establishment of assays for factors involved in mesoderm induction (reviewed by Slack, 1994). When animal caps are exposed to neuralizing factors, neural markers are expressed (Lamb et al., 1993). Thus, animal cap assays may be employed to study neural induction as well. In a popular adaptation of the traditional assays, animal caps are derived from embryos microinjected with RNA or DNA encoding potential inducing factors (Slack, 1994). In cases where a developmentally significant pathway is activated, an autoinductive diversion of cell fate may be promoted (Table 3).

Induction Assays in Whole Embryos. In whole embryos, the role of various putative inducing factors in axial patterning has been assessed using two principal criteria: the ability of ectopically expressed genes to elicit secondary axes, and the ability of overexpressed genes to rescue UV-irradiated embryos, which otherwise fail to display any dorso-ventral axis. Indeed, the neuralizing molecule noggin was identified using axial rescue of UV-irradiated embryos to assay pools of cloned DNAs derived from a complementary DNA (cDNA) library prepared from hyperdorsalized embryos (Smith and Harland, 1992). Lemaire et al. (1995) employed a similar strategy of expression cloning to isolate Siamois, a *Xenopus* homeobox gene that elicits a complete secondary axis when expressed from small quantities of mRNA injected into a single ventral blastomere of 4- or 8-cell embryos. In that study, axis duplication was chosen over axial rescue as the test parameter to avoid the experimental "noise" resulting from axial remnants present in 10–15% of UV-treated embryos.

OTHER GENE FUNCTION ASSAYS. Other laboratories are using embryo microinjections to study various other events in *Xenopus* development. Our laboratory, for example, has employed microinjected embryos to assess the role of cholinergic signaling in the development and maintenance of the vertebrate NMJ (Seidman et al., 1994,1995; Shapira et al., 1994). In a unique approach to the study of neurotransmitter release at developing NMJs, purified synapsin proteins were microinjected into blastomeres of cleaving embryos to obtain synapsin I- and IIa-loaded neurons. Spontaneous and evoked synaptic responses were then evaluated in NMJs formed in vitro from nerve–muscle cocultures (Lu et al., 1991; Schaeffer et al., 1994). In a later study, ultrastructural maturation of presynaptic nerve terminals was shown to accompany enhanced functional activity of synapsin-loaded NMJs (Valtorta et al., 1995). Similar studies were subsequently performed using in vitro-transcribed mRNA to overexpress the synaptic vesicle protein synaptophysin in microinjected embryos (Alder et al., 1995). Coinjection of rhodamine-conjugated dextran allowed identification of motor neurons overexpressing the heterologous protein. These neurons were then shown to display enhanced synaptic activity. Using Lucifer yellow to monitor gap junction activity, Olson et al. (1991) exploited microinjection techniques to study the effect of Wnt-related proteins on gap junctional communication in the early embryo. Using lineage tracers and timelapse photography, Niehrs et al. (1993) demonstrated the ability of the homeobox gene goosecoid to promote gastrulation-specific migration properties in vivo and in in vitro motility assays.

STUDYING GENE FUNCTION BY DOWNREGULATION

TARGETED mRNA DESTRUCTION. An alternative to heterologous overexpression in gene function studies is to examine the effects of downregulation of gene expression. A popular approach to the downregulation of specific gene expression in various experimental systems exploits antisense RNA- or oligonucleotide-mediated mRNA destruction. In light of technical limitations, however, RNA depletion assays using antisense RNAs or oligonucleotides have yet to enter into routine use in *Xenopus* embryos. Nevertheless, the development of chemically modified "protected" antisense oligonucleotides (reviewed by Eckstein, 1985) may permit advancement in this area. An additional, potentially promising technology for targeted mRNA degradation is the design of ribozymes (endonucleolytic RNAs that employ homologous base pairing to effect sequence-specific cleavage of other RNAs) (Tuschl et al., 1994). Unfortunately, ribozyme technology has yet to be demonstrated as effective in *Xenopus* (discussed by Vize et al., 1991).

Injection of Antibodies. Another approach to the downregulation of specific proteins is the microinjection of antibodies (Wright et al., 1989; Purcell et al., 1993). In the case of β-catenin (a component of the cadherin cell adhesion machinery), affinity-purified Fab fragments induced a secondary body axis when injected into ventral blastomeres (McCrea et al., 1993). In another example, antibodies against the synaptic vesicle-specific protein synaptophysin were shown to inhibit neurotransmitter secretion at NMJs (Alder et al., 1992). In those experiments, rabbit antifrog synaptophysin antibodies (0.05 mg/mL) were injected into a single blastomere of 2- or 4-cell embryos, which were cultured to stage 20–22. Nerve–muscle cultures were then prepared and subjected to electrophysiological analysis (*see* Peng et al., 1991 for a detailed protocol).

Dominant-Negative Molecules. For multimeric complexes, the creation of dominant-negative assemblies has proven to be an effective "knock out" strategy. In this approach, RNA encoding a biologically inactive mutant of one of the subunits is injected. The overexpressed but defective heterologous protein associates with its endogenous wild-type counterpart, titrating the native subunits into an inactive complex. The dominant-negative approach was successfully applied to disrupt the FGF (Amaya et al., 1991; MacNicol et al., 1993; Cornell and Kimelman, 1994; Schutte-Merker and Smith, 1995), activin (Hemmati-Brivanlou and Melton, 1994), and bone morphogenetic protein (BMP) (Graff et al., 1994) signaling pathways active in mesoderm induction. Dominant-negative expression studies have been particularly effective in revealing inhibitory signals in inductive pathways. For example, ectopic expression of mRNA encoding a truncated form of glycogen synthetase kinase 3 (GSK-3) induced complete secondary axes replete with head, notochord, and neural tube, when introduced into ventral blastomeres (He et al., 1995). Together with the observation that wild-type GSK-3 mRNA exerts a strong ventralizing effect when introduced into dorsal blastomeres, these studies led to the conclusion that GSK-3 inhibition is both necessary and sufficient to initiate dorsal axis specification.

HOST TRANSFER. To study the role of maternal RNAs in very early development, the host transfer technique has been applied (reviewed by Heasman et al., 1991). In this approach, oocytes are surgically removed from a mature female frog, labeled with a vital dye, microinjected, matured in vitro with progesterone, and transplanted into a host female, who is then hormone-stimulated to lay eggs. In vitro fertilization is performed on the dye-marked eggs. Utilizing this technology, maternal stores of EP-cadherin were depleted in blastula-stage embryos by microinjecting oocytes with partially protected antisense oligodeoxynucleotides targeted against EP-cadherin mRNA (Heasman et al., 1994). In that study, cell adhesion was severely disrupted both in whole embryos and in cell reaggregation assays. The mutant phenotype was rescued by microinjection of E-cadherin encoding mRNA into EP-cadherin mRNA-depleted oocytes.

Detection Strategies

The large variety of microinjection strategies in *Xenopus* embryos is favorably matched by the diverse analytical techniques that may be applied to assess their outcome. Gene expression is readily monitored on the levels of both RNA and protein. However, histology plays a large role in determining the developmental consequences of most *Xenopus* microinjection experiments. The types of analytical protocols commonly used with microinjected embryos are outlined in Table 4 and discussed in detail here.

Detection of RNA

Two complementary evaluations need be included in the assessment of any gene function assay: (1) determination of the level, timing, and distribution of both native and heterologous gene expression and (2) assessment of the developmental consequences of over-, under-, or ectopic gene expression. The high sensitivity of RNA detection methods often makes this the preferred level of analysis. Standard procedures for the extraction and analysis of RNA are readily adaptable to *Xenopus* embryos and include RNA (Northern) blotting (Christian and Moon, 1993), RNase protection (S1) analysis (Ferreiro et al., 1994; MacNicol et al., 1994), and RT-PCR (Ben Aziz-Aloya et al., 1993a; Seidman et al., 1995; Witta et al., 1995). In most cases, the equivalent of only 1–2 embryos is required. The quantitation of RNA may be used to assess the transcriptional activity of an injected DNA, the stability of an injected message, or the activation of endogenous *Xenopus* genes by heterologous gene expression (Hopwood and Gurdon, 1990). Analyses may be performed on whole embryo extracts (Smith and Harland, 1991), on manually dissected embryos (Jonas et al., 1987; Godsave et al., 1994), or on animal caps (Amaya et al., 1991; Ferreiro et al., 1994).

Detection of Proteins

On the level of proteins, low yields and the lack of sensitive assays for many of the proteins involved in early

Table 4
Analytical Techniques in *Xenopus* Embryos

Level of analysis	Applicable protocols	When useful
Molecular[a]	Northern blots RT-PCR S1 nuclease protection	Monitor transcription from injected plasmids Detect induction of endogenous gene expression
Biochemical[b]	Enzyme activity assays Biosynthetic labeling and SDS-PAGE Western blots Receptor binding assays	Detect overexpression of heterologous protein Detect induction of endogenous gene expression
Histochemical[c]	General morphology Ultrastructural analyses *In situ* hybridization Immunocytochemical staining Cytochemical activity staining	Assay misexpressed genes for developmental or morphogenetic functions Detect cell or tissue-specific gene expression Examine determinants of cellular/subcellular localization of heterologous RNA/protein
Functional[d]	Scoring for developmental defects/changes in cell fate Morphometric measurements Electrophysiological recording Video recording Cell adhesion assays	Assay misexpressed proteins for developmental or morphogenetic functions For studies of synaptogenesis For studies of cell adhesion, cell migration

[a]RNA is readily prepared from whole or dissected embryos.

[b]Biochemical analyses may be performed on whole-cell extracts or subcellular fractions.

[c]Histochemical analyses are routinely performed on whole-mount embryos or in thin sections and examined by light, fluorescence, confocal, and electron microscopy.

[d]Assays for biological function are performed on whole embryos, on dissociated cells, and on cultured neurons or muscle cells.

development often preclude their accurate quantification in microinjected embryos. In that case, cytochemical staining is generally the detection method of choice. Nonetheless, expression of processed Vg-1 was demonstrated in stage 7, mRNA-injected embryos by denaturing gel electrophoresis following in vivo labeling of injected eggs with a mixture of ^{35}S-labeled amino acids (0.3 μCi/embryo; Thomsen and Melton, 1993). However, radioactive labeling of injected embryos may not be suitable for analysis at later stages, as the label appears to interfere with gastrulation (Seidman, unpublished observations). In the case of human AChE expressed from the CMV promoter, immunoblotting with a species-specific monoclonal antibody (MAb) was used to visualize the heterologous enzyme (Seidman et al., 1994). However, in that study, homogenates of 200 gastrula-stage embryos were required and subjected to immunoaffinity purification before a detectable signal could be achieved.* In contrast, a sensitive colorimetric activity assay revealed 50–100-fold overexpression of AChE compared to control uninjected embryos using the equivalent of only 1/2–1/3 embryo/assay (Seidman et al., 1994). Graff et al. (1994) successfully utilized epitope tags and immunoblotting to monitor Raf-1 expression in protein lysates from microinjected embryos.

Histology

Good histology and photomicroscopy are indispensable in *Xenopus* work (reviewed by Kelly et al., 1991). Both *in situ* hybridization and immunocytochemistry are routinely employed on sectioned tissues and in whole mount (Dent et al., 1989; Hemmati-Brivanlou et al., 1990; Harland, 1991; Klymkowsky, 1991; O'keefe et al., 1991). The advan-

*We have recently achieved good results in immunoblots prepared from crude embryo extracts using an enhanced chemiluminescence (ECL) detection protocol (Sternfeld et al., unpublished data).

tage of whole-mount techniques is that they permit relatively rapid screening of all embryonic tissues from large numbers of embryos—an almost insurmountable task to perform on sectioned embryos (although whole-mount embryos can be subsequently sectioned for high magnification light microscopy). The production of publication-quality photographs of whole-mount embryos requires time and patience. However, good pictorial documentation is essential and worth the effort. Powerful quantitative analysis of "serial sections" from whole-mount immunostained embryos may be obtained using confocal laser microscopy (Cary and Klymkowsky, 1995; *see also* Beumer et al., 1995 for technical tips on the use of fluorescent dyes in *Xenopus*), where this technologically advanced equipment is available. *Xenopus* embryos are also amenable to electron microscopy analysis (Billet and Gould, 1971; Kelly et al., 1991; Cary and Klymkowsky, 1995; *see also* Experimental Methodologies in Chapter 2). Cytochemical activity staining and electron microscopy were applied to the examination of developing muscle and NMJs in 2- and 3-d-old embryos overexpressing human AChE (Seidman et al., 1994) or murine AChR (Shapira et al., 1994). In those studies, the synaptic accumulation of overexpressed cholinergic proteins was correlated with changes in NMJ ultrastructure.

Artifacts

Among the commonly observed developmental defects in batches of microinjected embryos are abortive gastrulation, kinking of tailbud stage embryos, and death (Vize et al., 1991). However, these perturbations in normal development may be attributable to initial poor health of the eggs or microinjection damage. Since these defects occur with varying frequencies in both uninjected and sham-injected embryos, it is important to score their number and nature in each experiment and compare with appropriate controls.

The Vertebrate Neuromuscular Junction

The vertebrate NMJ is a highly specialized interface that facilitates rapid and efficient communication between motor neuron and muscle. The NMJ is a cholinergic synapse: ACh released from the nerve terminal diffuses across the synaptic cleft where it interacts with the nAChR to stimulate a short ion flux and depolarization of the postsynaptic membrane. Released from the receptor, ACh is rapidly hydrolyzed by AChE to acetate and choline, terminating the signal. Choline is at least partially recycled via a presynaptic choline uptake pathway and used as a substrate for ChAT to regenerate presynaptic ACh stores.

Formation of a functional NMJ requires the targeted deposition of numerous synaptic proteins at the nerve–muscle contact site (Flucher and Daniels, 1989; Froehner, 1991; Ohlendieck et al., 1991). However, it is not yet clear how the selective accumulation of synaptic proteins at the NMJ is accomplished. In the case of AChR and AChE, accumulated evidence implicates transcriptional, translational, and post-translational mechanisms (Changeux, 1991). Moreover, the morphological and biochemical differentiation of the NMJ appears to involve a carefully orchestrated interplay between the muscle and the nerve, which could, in principle, invoke cholinergic signaling pathways (Dan and Poo, 1994).

Neuromuscular Junction Structure

The most striking ultrastructural feature of a mature NMJ in vertebrate skeletal muscle is the deep invaginations of the postsynaptic cell membrane, which form the junctional folds (Figs. 11 and 12). The crests of the folds, which appear as thick electron-dense domains, are highly enriched in immobilized AChR (10,000–12,000 mol/μm^2; Salpeter and Loring, 1985) and are positioned directly opposite presynaptic "active zones" enriched in ACh-releasing vesicles. In the bottom two-thirds of the fold, the concentration of AChR drops 10–100-fold, and within a few microns outside the junction, "nonjunctional" receptor

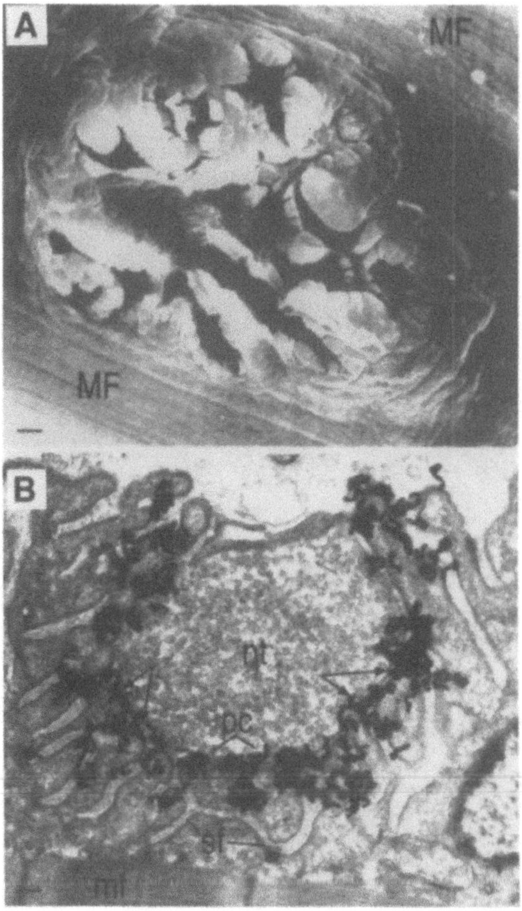

Fig. 11. Electron microscope view of the vertebrate NMJ. **(A)** Scanning electron micrograph of lizard intercostal muscle NMJ. In this photograph, the nerve has been removed to reveal topological features of the intact postsynaptic membrane. The oval shape of the lizard NMJ is similar to that of the mammalian junction (magnification ×4000). Size bar, 4 μm. Reprinted from Bartol et al. (1991) with permission from *Biophysical Journal*. **(B)** Transmission electron micrograph of a cross-section from a lizard NMJ. Note localization of AChR (arrows) at the crests of the juctional folds, in close apposition to the presynaptic nerve terminal. mf, myofibrils; nt, nerve terminal; pc, primary cleft; sf, secondary fold, ACh-containing synaptic vesicle (magnification ×11,000). Size bar, 0.4 μm. Adapted from Salpeter (1987). Photographs courtesy of Miriam Salpeter.

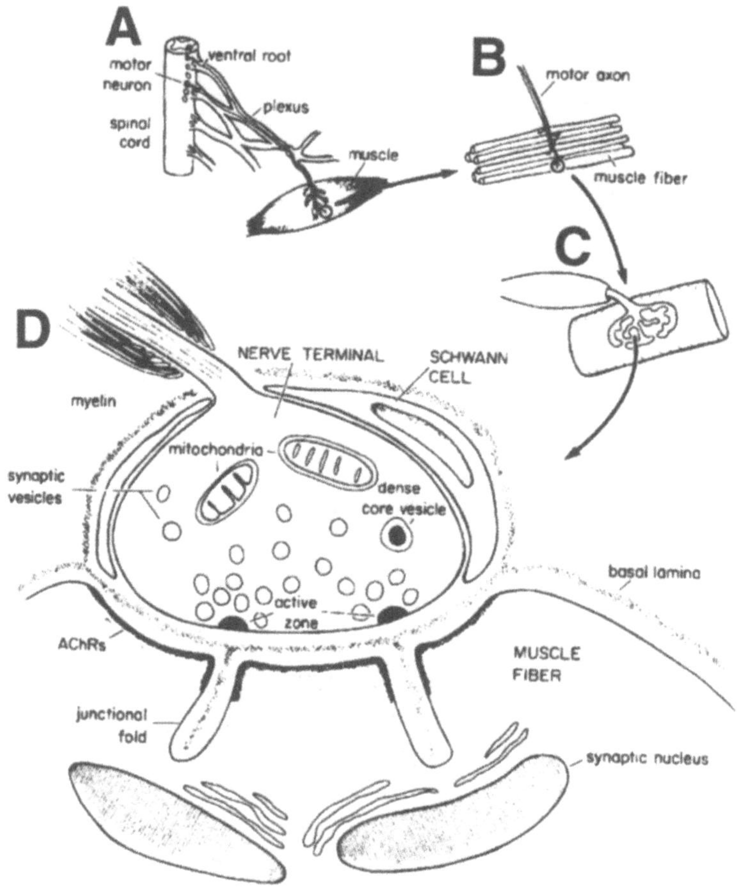

Fig. 12. Schematic view of the vertebrate NMJ. **(A)** Motor neurons in the ventral horn of the spinal cord send axons through peripheral nerves to innervate muscles, usually in a central innervation band. **(B)** Within the muscle, axons branch to innervate multiple muscle fibers, with each fiber receiving a single synaptic input. **(C)** The axonal branch loses its myelin sheath to terminate in a spray of boutons on the muscle fiber surface. **(D)** Electron microscope view of the neuromuscular synapse. Reprinted from Hall and Sanes (1993) with permission from Cell Press.

densities plummet to <10 mol AChR/μm^2. Other proteins associated with the postsynaptic membrane include voltage-sensitive sodium channels, β-1 integrins, and

N-CAM. In addition, an impressive list of presynaptic and postsynaptic cytoskeletal proteins, including dystrophin, dystrophin-related protein (DRP/utrophin), the 43 kDa peptide (rapsyn), syntrophin, dystroglycan, and adhalin are also concentrated at the NMJ and likely provide the infrastructure necessary to maintain the structural and organizational integrity of the junction (Froehner, 1991; Hall and Sanes, 1993; Apel and Merlie, 1995).

The muscle fiber is encased by a specialized extracellular matrix (ECM) called the basal lamina, which fills the 50-nm-wide synaptic cleft (Hall and Sanes, 1993). The basal lamina is highly enriched in AChE (Hall, 1973; McMahan et al., 1978). AChE is associated with the basal lamina through ionic interactions that probably involve heparan sulfate proteoglycan (Brandan et al., 1985; Inestrosa and Perelman, 1989). Other proteins associated with the basal lamina include agrin, collagen, laminin A, and S-laminin (Carbonetto and Lindenbaum, 1995). Following experimentally induced degeneration of both nerve and muscle from an existing synapse, the basal lamina retains indelible cues that attract the reinervating axon toward the original synaptic site (Anglister and McMahan, 1985). This synaptic imprint also appears to include factors that are able to promote synapse-specific gene transcription (*see* Synapse-Specific Transcription of Synaptic Proteins section in this chapter). The leucine-arginine-glutamate (LRE) tripeptide (Hunter et al., 1989), present in several synaptic basal lamina proteins, including AChE and laminin, may play a role in mediating interactions between the incoming axon and the basal lamina (Hall, 1995; Porter et al., 1995).

All together, the finely integrated communications network mediating neurotransmission across the NMJ has prompted the suggestion that the entire junction be viewed as a single superstructure that extends from the cytological specializations of the presynaptic nerve terminal to the postsynaptic junctional nuclei, and includes everything in between (Hall and Sanes, 1993).

Aggregation of Acetylcholine Receptor/Acetylcholinesterase

Myoblasts, the undifferentiated predecessors of the muscle fiber, express low levels of diffuse, nonaggregated AChR on their membranes. Following fusion, both AChR and AChE begin to appear on the cell surface, evenly distributed. Although AChR will spontaneously form clusters in the absence of neural input, neural encroachment on the myofiber can stimulate AChR clustering directly beneath the approaching axon within a few hours. At least some of the AChR appearing in these clusters can be attributed to the lateral migration and aggregation of preexisting receptor molecules (Anderson and Cohen, 1977; Anderson et al., 1977; reviewed by Apel and Merlie, 1995). At the synapse, AChR clustering is associated with a 43-kDa protein and a complex of dystrophin/utrophin-associated transmembrane glycoproteins (dystrophin–glycoprotein complex [DGC]) that appear to coordinate anchoring of the receptor to the cytoskeleton (Apel et al., 1995).

An AChR clustering activity was first shown to be present in extracts of *Torpedo* electric organ (Godfrey et al., 1984). This activity was subsequently purified and named agrin (Nitkin et al., 1987). The observations that purified agrin induces aggregation of AChR in cultured myotubes without affecting the total number of surface-associated receptor molecules, and that antiagrin antibodies block nerve-induced AChR aggregation, strengthened the contention that agrin plays a prominent role in AChR clustering in vivo (reviewed by Wallace, 1991; Nastuk and Fallon, 1993). Moreover, two observations suggested that agrin stimulates a cascade of events associated with postsynaptic differentiation rather than merely serving as a structural component of AChR clusters: the nonstoichiometric effects of agrin (e.g., agrin:AChR = 1:200; Nitkin et al., 1987), and the fact that AChE and other postsynaptic elements are colocalized with AChR in agrin-mediated clusters. Indeed, several lines of evidence suggest that the AChR-aggregating activity of agrin is mediated, at least in part, by tyrosine kinase phosphorylation of the β-AChR subunit (Wallace, 1991,1994; Baker and Peng, 1993).

Studies on agrin isoforms derived from alternative splicing implicated nerve-specific transcripts in AChR aggregating activity and led to the conclusion that agrin may participate in synaptogenesis throughout the nervous system (reviewed by McMahan et al., 1992). Motoneuron-specific isoforms of agrin released into the synaptic cleft become embedded within the basal lamina. Although muscle cells also synthesize agrin (Lieth et al., 1992), AChR clustering is dependent on the neuron-derived pool (Reist et al., 1992). In cultured muscle cells, components of the DGC colocalize with AChR (Campanelli et al., 1994; Apel et al., 1995). At least one of these, α-dystroglycan, binds agrin. A MAb directed against α-dystroglycan blocked the binding of agrin in vitro and inhibited agrin-induced AChR aggregation in S27 cells, suggesting that α-dystroglycan may serve as a functional agrin receptor (Gee et al., 1994). However, subsequent data indicated that the crucial component in agrin-mediated receptor aggregation is a muscle-specific receptor tyrosine kinase (MuSK) complex (Glass et al., 1996). Targeted genomic disruption of either agrin (Gautam et al., 1996) or MuSK (DeChiara et al., 1996) in mice results in markedly impaired NMJ formation during embryogenesis.

One prominent structural component of AChR clusters is a cytoskeletal protein designated the 43-kDa protein. The 43-kDa protein is colocalized with AChR in clusters and in NMJs in a 1:1 ratio, and alkaline extraction of peripheral membrane proteins, including the 43-kDa protein, results in dispersal of local AChR accumulations (reviewed by Froehner, 1991). In microinjected *Xenopus* oocytes, coexpression of AChR and the 43-kDa protein induced clustering of heterologous AChR on the oocyte membrane, whereas expression of AChR alone promoted uniform distribution of the exogenous receptor (Froehner et al., 1990). Similar results were obtained using transfected fibroblast cells (Phillips et al., 1991). Together, these observations indicate that the 43-kDa protein is directly involved in the immobilization of AChR in clusters. Perhaps for these reasons, an AChR-associated protein detected in rapid freezing cryoelectron microscopy studies was presumed to represent the 43 kDa protein (Sansom, 1993). In order to distinguish the AChR-associated 43-kDa protein from other 43-kDa components of the postsynaptic membrane, the name rapsyn ("receptor-associated protein in the synapse") was proposed (Apel et al., 1995). Following the observation that rapsyn can independently recruit both AChR and dystroglycan (but not other membrane-associated proteins) into clusters in transfected fibroblast cells, rapsyn was suggested to function as a structural mediator, physically linking AChR to the DGC (Apel et al., 1995; *see also* Gautam et al., 1995).

Synapse-Specific Transcription of Synaptic Proteins

The global increase in AChR molecules on the surface of differentiated myotubes is not dependent on innervation or neuromuscular activity, and appears to result from enhanced transcription of the genes encoding the individual AChR subunits (Buananno and Merlie, 1986). Therefore, upregulation of AChR expression is probably part of the overall myogenic differentiation program. However, during maturation of the muscle fibers, AChR subunit-encoding mRNAs become selectively localized to the junctional nuclei under the end plate (reviewed by Changeux, 1991; Jennings and Burden, 1993). Compartmentalization of AChR mRNAs within the junctional domain appears to result from two complementary processes: selective transcription of AChR-encoding mRNAs in junctional nuclei and selective repression of transcription in extrajunctional sites.

Nerves appear to secrete a soluble factor capable of selectively stimulating transcription of AChR-encoding mRNAs in the junctional nuclei (Hall and Sanes, 1993). The observation that AChR subunit mRNAs continue to be expressed for several weeks following denervation and that regenerating muscle transcribes AChR mRNA in nuclei underlying the original synaptic sites implies that the stimulus provided by the nerve promotes a relatively long-lasting effect on regulation of transcription in junctional nuclei, and that the basal lamina carries a durable imprint of the initial presence of the nerve. Two molecules were long considered candidates to participate in synapse-specific regulation of transcription: acetylcholine receptor-inducing activity (ARIA) and calcitonin gene-related protein (CGRP). Both ARIA and CGRP stimulate accumulation of AChR-subunit-mRNA when added to cultured muscle cells. CGRP is synthesized by motor neurons and transported to the nerve terminals from where it is released upon stimulation. ARIA is a member of the neuregulin (NRG) family of ErbB2/Neu receptor tyrosine kinase ligand activators (Carraway and Burden, 1995). ARIA is present in brain and spinal cord. Motoneuron-derived NRGs were shown to be concentrated at NMJs and to activate AChR expression in a manner dependent on the 181-bp response element shown to mediate synapse-specific, electrical activity-dependent transcription of the δ-AChR subunit (Jo et al., 1995). Agrin does not display AChR gene regulating activity.

The selective loss of extrajunctional AChR has been clearly associated with electrical, but not mechanical, activity of the muscle (Changeux et al., 1991; Hall and Sanes, 1993; Bessereau et al., 1994). Thus, agents blocking electrical muscle activity stimulate extrajunctional AChR expression, whereas chronic stimulation may repress AChR biosynthesis. This process appears to be calcium-dependent, probably involves second messengers, and may work through protein kinase C (Mendolzon et al., 1994). Both selective activation of junctional nuclei and selective repression of extrajunctional nuclei with regard to AChR-subunit-encoding genes have been demonstrated in transgenic mice

(Merlie and Kornhauser, 1989; Cahine et al., 1992; Jo et al., 1995).

Cholinergic Signaling and Neuromuscular Pathologies

Unimpaired neurotransmission across the NMJ is essential for normal motor function. Thus, acute poisoning of cholinergic signaling pathways may lead to paralysis and death. Not surprisingly, therefore, numerous natural and synthetic neurotoxins are targeted against either AChR or AChE (Soreq and Zakut, 1993). Clinically, patients suffering chronic autoimmune insults to key elements of the cholinergic system display muscle weakness, fatigue, and a variety of related secondary symptoms (*see* Chao et al., 1982 for ChAT; Livneh et al., 1988 for AChE; and Schonbeck et al., 1990 for AChR). In some patients with congenital myasthenia, ultrastructural and physiological alterations of NMJs were associated with AChE and/or AChR deficiencies (Wokke et al., 1989; Jennekens et al., 1992). In one of these syndromes, patients presented, in addition to AChE/AChR deficits, NMJs displaying decreased miniature end-plate potentials, reduced postsynaptic membrane lengths, and severely impaired postsynaptic secondary folding (Smit et al., 1988). In one case, downregulation of AChR appeared to accompany loss of synaptic AChE (Jennekens et al., 1992). It is noteworthy that chronic exposure to anti-cholinesterase (ChE) drugs may induce similar myopathic symptoms (*see* Neuromuscular Junction Acetylcholinesterase section in this chapter). Thus, balanced cholinergic neurotransmission may play a crucial role in NMJ development and maintenance.

To study vertebrate NMJ formation and function, it is necessary to find an accessible experimental model—one which is amenable to molecular biological and biochemical as well as electrophysiological and ultrastructural analyses. The amphibian model, and particularly the South African clawed frog *Xenopus laevis*, offers these features. In addition, one necessarily needs to choose an appropriate model synaptic molecule to study in the hope of gleaning general principles regarding the complex organization and physiological interactions among synaptic proteins. For this purpose, we have selected AChE as our model.

Acetylcholinesterase

The ACh hydrolyzing enzyme, AChE, has long been noted for its vital role in mitigating and terminating neurotransmission at cholinergic synapses (Dale, 1914; Loewi and Navratil, 1926). However, with the advent of sensitive biochemical assays (Ellman et al., 1961; Johnson and Russell, 1975) and histochemical staining techniques (Koelle and Friedenwald, 1949; Karnovsky and Roots, 1964), ChE activities, including those of AChE and the related enzyme BuChE, were revealed in a variety of what would otherwise be considered noncholinergic cells and tissues. Thus, AChE expression has been observed not only in muscle (Hall, 1973) and in adult brain (Greenfield, 1984; Gennari and Brodbeck, 1985), but also in embryonic neurons not destined to function as cholinergic neurons (Robertson, 1993), hematopoietic cells (Alles and Hawes, 1940; Paulus et al., 1981), germ cells (Nelson, 1964; Gundersen and Miledi, 1983; Seidman, unpublished observations), and various embryonic cells and tissues (Drews, 1975; Fitzpatrick-McElligot and Stent, 1981; Layer and Willbold, 1995). Furthermore, amplification and aberrant expression of the acetylcholinesterase gene (ACHE) encoding human AChE has been associated with a variety of neoplastic tissues and tumor cell lines (Zakut et al., 1988,1990; Soreq and Zakut, 1990a,b,1993).

The role of ChEs in noncholinergic, nonnervous cell types is yet unknown. However, the spatio-temporal patterns of ChE expression observed in some tissues may imply a role for ChEs in cell proliferation, differentiation, or morphogenesis (Drews, 1975; Soreq et al., 1994; Layer and Willbold, 1995). Moreover, the presence of ChE activities in morphogenetically active biological systems may imply a role for ACh and cholinergic signaling pathways in development (Rama Sastry and Sadavongvivad, 1979). Sequence homologies between AChE and the neuronal cell-surface proteins neurotactin (de la Escalera et al., 1990), gliotactin (Auid et al., 1995), and neuroligin (Ichtchenko et al., 1995) could predict a role for AChE in heterologous cell–cell recognition processes. Several studies have indeed demonstrated developmental roles for AChE that appear to function independently of its hydrolytic activity (Layer et al., 1993; Small et al., 1995). The molecular and cellular biology of AChE have recently been comprehensively reviewed (Massoulie et al., 1993; Soreq and Zakut, 1993; Taylor and Radic, 1994). Therefore, the following discussion will be limited to those aspects directly related to our studies in microinjected *Xenopus* embryos.

Biological Roles

Neuromuscular Junction Acetylcholinesterase

AChE is highly enriched in the basal lamina of the vertebrate NMJ, where it participates in termination of the postsynaptic signal. Experimental and theoretical models have been established to examine the precise role of AChE in NMJ neurotransmission (Katz and Miledi, 1973; Rosenberry, 1979; Wathney et al., 1979; Land et al., 1980,1981, 1984; Bartol et al., 1991). Quantal packets of ACh (approx 10,000 mol/quanta) released from the presynaptic nerve terminal create high local concentrations of neurotransmitter, which probably saturate synaptic AChE, allowing the majority of ACh molecules to traverse the cleft and interact with postsynaptic AChRs to evoke membrane depolarization. Unbinding from its receptor, ACh is rapidly hydrolyzed by AChE and prevented from initiating additional binding events. Inhibition of AChE appears to prolong the falling phase (t_f) of miniature end-plate currents (mEPCs) while exhibiting little effect on either the rise time (t_r) or number of channels open at the peak (A_c). Under conditions of complete AChE inactivation, t_f is observed to approach AChR channel closing time. Therefore, the primary role of AChE in the NMJ is likely the prevention of "buffered diffusion" of ACh within the synapse (i.e., persistence of ACh within the synaptic cleft that results in multiple binding events and prolonged mEPCs). In vivo, protracted postsynaptic currents may disrupt normal 1:1 firing and result in muscle spasms and/or paralysis. Long-term inhibition of AChE may induce pathologic changes in NMJ ultrastructure and electrical activity (Engel et al., 1973; Laskowski et al., 1975; Glazer et al., 1978). The myopathic consequences of chronic AChE inhibition have been attributed to an excess in intracellular Ca^{2+} resulting from prolonged membrane depolarization (Leonard and Salpeter, 1978).

Acetylcholinesterase in the Central Nervous System

Histochemical studies place catalytically active AChE in basal brain nuclei, hippocampus, brain stem, and specific regions of the cortex (Landwehrmeyer et al., 1993; Beeri et al., 1995). In cholinergic neurons of the CNS, AChE is presumed to play its familiar role of mitigating ACh-mediated neurotransmission (Soreq and Zakut, 1993). However, the presence of AChE in nonneuronal brain cells (Ord and Thompson, 1953), otherwise noncholinergic regions of the brain (Greenfield, 1984), and in developing neurons prior to the acquisition of ACh sensitivity (Layer and Willbold, 1995) drives the search for alternative roles for AChE (*see* Chapter 4).

Cholinergic deficits are associated with several neurodegenerative conditions (reviewed by Schwarz et al., 1995a). In theses cases, AChE activity may be altered. For example, in Alzheimer's disease (AD), a significant reduction in the G4/G1 ratio was attributed to selective loss of

G4 from presynaptic cholinergic structures (Rakonczay and Brimijoin, 1988). Aberrant expression of AChE has also been associated with Parkinson's disease, suggesting that cholinergic dysfunction may accompany dopaminergic deficits in this neurodegenerative disease as well. Changes in the levels of soluble G4 AChE in cerebrospinal fluid may be related to several other neurodegenerative pathologies involving cholinergic deficits (Soreq and Zakut, 1993). Persistently high levels of G4 in amniotic fluid is indicative of neural tube defects in the developing embryo (Bonham et al., 1987).

Cholinergic signaling in the CNS has been associated with cognitive functions, including short-term learning and memory (Kadar et al., 1990 and references therein). In this light, the selective loss of cholinergic neurons in the brains of AD patients led to the cholinergic theory of AD. The cholinergic theory of AD proposes that loss of cholinergic neurons leads to an ACh deficit, which is correlated with the progressive deterioration of central cholinergic systems and the accompanying degeneration of cognitive function characteristic of AD patients (Katzman, 1986; Coyle et al., 1983; Davis et al., 1993). A widely accepted approach to the retardation of AD-related demise of cognitive mental functions has therefore been to counter the ACh deficit with ChE inhibitors (Davis et al., 1993). Indeed, Tacrine (tetrahydroaminoacridine [THA], Cognex)—the first FDA-approved Alzheimer's drug—is a ChE blocker that displays at least limited efficacy in ameliorating the decline of some cognitive functions in early-stage AD patients (Knapp et al., 1994).

Embryonic Acetylcholinesterase

The embryonic expression of AChE has been rigorously reviewed by Drews (1975), following a detailed histochemical investigation. In that study, AChE activity was correlated to morphogenetic movements during organogenesis. Both cells that would eventually express AChE routinely, and those in which AChE expression was evident only during morphogenic stages, were observed. AChE expressed during organogenesis was operationally defined as embryonic ChE. Embryonic ChE activity was observed during gastrulation and neurulation, and during development of the head, heart, limb bud, pancreas, and liver. In addition, expression of AChE during migration of individual cells such as neural crest and primordial germ cells was noted. In all these cases, microfilament-dependent, pseudopodial, morphogenetic movements were distinguished from other forms of cell motility and microtubule-dependent processes. Furthermore, these cellular movements were noted to involve calcium fluxes and alterations in cell adhesion properties. In other studies, evidence for the involvement of cholinergic signaling in ciliary function was noted; examples include unicellular organisms, mammalian spermatozoa, and ciliated epithelial tissues (Rama Sastry and Sadavongvivad, 1979).

Prior to synaptogenesis, both AChE and BuChE may be expressed in migrating and differentiating neurons (Layer and Willbold, 1995). In the developing avian nervous system, BuChE expression is associated with migratory, mitotically active cells, whereas the appearance of AChE signals the cessation of mitosis and the onset of terminal differentiation (Layer and Willbold, 1995). A close temporal relation was further observed between AChE expression and projection of neurites; AChE-positive cells in the neural tube were invariably those destined to project long efferent axons, including motor axons innervating the developing myotomes. In the somites, a rostral-caudal wave of BuChE expression was followed by the appearance of AChE. However, BuChE-rich areas were those through which neurite extensions penetrated into the myotomes. Overall, the regular, sequential, alternating expression of AChE and BuChE in the chick nervous system suggested coordinated regulation of the two genes, but implied distinct functions for the two enzymes.

Hematopoietic Acetylcholinesterase

Both erythrocytes and megakaryocytes, generally assumed to share a common progenitor, display conspicuous expression of AChE (Soreq et al., 1994). Thus, mammalian erythrocyte membranes have proven to be a rich source of AChE for experimentation, and secreted AChE has been exploited as a marker of megakaryocyte differentiation in culture (Burstein et al., 1985). Although antisense inhibition studies of cultured bone marrow cells indicate a role for ChEs (AChE and BuChE) in proliferation and differentiation of erythroid precursors (Patinkin et al., 1990; Lapidot-Lifson et al., 1992; Lev-Lehman et al., 1994; Soreq et al., 1994), the precise role and mechanism of their involvement in blood cell development is yet unknown. Following molecular cloning of the human ACHE encoding AChE (Soreq et al., 1990), it was discovered that a His to Asp substitution at position 322 in human AChE is responsible for the allelic variation in the YT blood type (Bartels et al., 1993).

Megakaryocytes are the source of circulating blood platelets, and both display AChE activity (Chuang, 1973). ACh analogs and ChE inhibitors promote megakaryocytopoiesis in culture (Burstein et al., 1980) and in vivo (Burstein and Harker, 1983). Moreover, amplifications of the genes encoding AChE (ACHE) and BuChE (butyrylcholinesterase gene [BCHE]), and breakages at or around the chromosomal loci harboring these genes (7q22 and 3q26, respectively) have been correlated with several platelet-count disorders and myeloid leukemias (Lapidot-Lifson et al., 1989). In this context, it is not surprising to discover that chronic occupational exposure to organophosphorous insecticides has been correlated with an increased risk of leukemia (Brown et al., 1990).

Acetylcholinesterase Gene

The genes encoding AChE from evolutionarily diverse species have been cloned and sequenced, including *Drosophila* (Hall and Spierer, 1986), *Torpedo* electric fish (Schu-

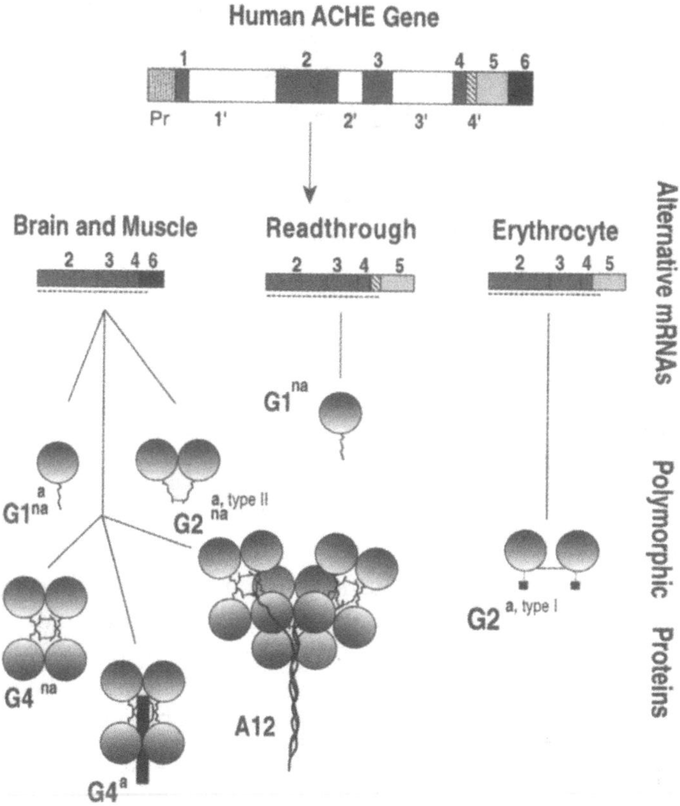

Fig. 13. The human ACHE gene *(full caption continues on p. 62)*.

macher et al., 1986; Sikorav et al., 1987,1988; Maulet et al., 1990), mouse (Rachinsky et al., 1990), rat (Legay et al., 1993b), and human (Soreq et al., 1990). The human ACHE, comprised of 6 exons and 4 introns (Fig. 13), spans at least 7 kb and maps to chromosomal position 7q22 (Ehrlich et al., 1992). A functional promoter lies within a 596-bp upstream sequence (Ben Aziz-Aloya et al., 1993a) that contains consensus recognition signals for the transcription factors MyoD, ATF/CREB, Sp1, E-box, Zeste, AP-2, GAGA, and EGR1 (Ben Aziz-Aloya et al., 1993b) (Fig. 14). Additional promoter elements may be located further upstream (Li et al., 1991; Mutero et al., 1995). Little is known, however, about in vivo, tissue-specific regulation of transcription of the ACHE (Fuentes and Taylor, 1993; Mutero et al., 1995).

Exons E2–E4 constitute the invariant coding sequence and contain all the information necessary to produce a catalytically active protein. Alternative usage of the 3' exons E5 and E6 could be expected, by homology to other AChE-encoding genes (Sikorav et al., 1988; Li et al., 1993; Legay et al., 1993b), to generate AChEs differing in their hydrodynamic properties and modes of cell association (*see* Acetylcholinesterase—The Enzyme section in this chapter). In addition, alternative usage of pseudointron I4 has been predicted to yet a third alternative AChE subtype (Karpel et al., 1994,1995). Three potential sites for N-linked glycosylation are presenting human AChE, but the extent

---

Fig. 13. *(previous page)* The human ACHE gene. The human ACHE gene encoding AChE, its alternative mRNAs, and polymorphic protein products are schematically presented. For a detailed presentation of the upstream promoter region (Pr), *see* Fig. 14. The common exons E2–E4 are shaded dark gray and the obligatory introns are shaded white. The pseudointron I4 is hatched. Alternative exons encoding variable C-terminal peptides are indicated: E5 (light gray) and E6 (black). The three alternative ACHE mRNAs are depicted with the open reading frames indicated by dotted lines. On the protein level, the various molecular configurations of the AChE subtypes are schematically presented: Shaded circles represent the common sequence encoded by exons E2–E4. Brain and muscle forms: Twisted thin lines represent E6 peptide, and dotted line represents putative readthrough peptide. $G_2^{a/na}$ represent disulfide-linked dimers of the nonamphiphilic (na) and type II amphiphilic (a) types. $G_4^{a/na}$ represent the DS (a) and LSS (na) tetramers formed by noncovalent pairing of dimers. G_4^{a} includes a 20-kDa hydrophobic noncatalytic subunit. A_{12} represents a high-mol-wt complex composed of 3 tetramers, each covalently linked to one strand of the triple-helical Q ("tail") subunit. Readthrough form: Readthrough ACHE mRNA gives rise to a monomeric AChE species with hydrophilic properties in transfected cells. Erythrocyte form: G_2^{a}: Type I amphiphilic dimer found on vertebrate erythrocyte membranes and *Torpedo* electric organ. Following endoproteolytic cleavage of the hydrophobic E5-peptide, the glycolipid phosphatidylinositol anchor (box) is attached to the C-terminal amino acid.

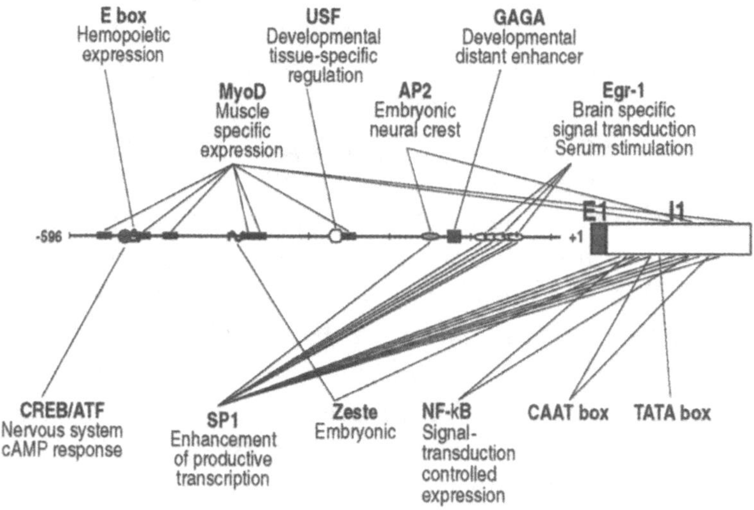

Fig. 14. The human ACHE gene upstream region. Schematic presentation (not to scale) of the 2.2-kb upstream region of the human ACHE gene including a 596-bp proximal promoter. The start site of transcription is marked +1. Note the numerous and varied transcription factor recognition sites suggesting tissue-specific regulation of transcription. This entire domain, but not 5'-truncated versions, was sufficient to direct transcription in microinjected *Xenopus* oocytes and embryos (Fig. 22). Reprinted from Ben Aziz-Aloya et al. (1993b) with permission from National Academy of Sciences.

and nature of glycosylation may vary in different tissues (Meflah et al., 1984; Liao et al., 1991,1992). The high G,C content (64%) comprising the human AChE coding sequence is characteristic of housekeeping genes expressed early in the cell cycle and may predict high thermal stabilities and extended half-lives for both AChE mRNA and its polypeptide product (Ben Aziz-Aloya, 1994).

Acetylcholinesterase Gene mRNAs

AChE is subject to tissue-specific and developmentally regulated posttranslational processing, which gives rise to a complex array of molecular forms varying in their extents of oligomeric assembly, association with noncatalytic subunits, hydrodynamic properties, and sites of subcellular localization (Fig. 13; *see* Massoulie et al., 1993; *see also* Acetylcholinesterase—The Enzyme section in this chapter). The principal level at which this diversity is controlled appears to be alternative splicing of 3' exons (Sikorav et al., 1988; Li et al., 1991; Karpel et al., 1994). Three alternative AChE-encoding mRNAs have been described in mammals. The dominant brain and muscle AChE found in NMJs (AChE-T) is encoded by an mRNA carrying exon E1 and the invariant coding exons E2, E3, and E4 spliced to alternative exon E6 (Soreq et al., 1990; Li et al., 1991; Ben Aziz-Aloya et al., 1993a). AChE mRNA bearing exons E1–E4 and alternative exon E5 encodes the glycolipid phosphatidylinositol (GPI)-linked form of AChE, characteristic of vertebrate erythrocytes (AChE-H; Legay et al., 1993a; Li et al., 1993). An additional readthrough mRNA species retaining the pseudointronic sequence I4 located immediately 3' to exon E4 was reported in rodent bone marrow and erythroleukemic cells (Rachinsky et al., 1990; Li et al., 1991; Legay et al., 1993a) and in various tumor cell lines of human origin (Karpel et al., 1994). A 3′ stop codon in I4 predicts a 26 amino acid hydrophilic C-terminus for this enzyme form.

Acetylcholinesterase—The Enzyme

AChE rapidly hydrolyzes ACh into choline and acetate. Its extraordinarily high catalytic turnover rate of 10^4 mol ACh/s (Rosenberry, 1975) makes AChE one of the fastest enzymes known and suggests that diffusion of substrate is probably rate limiting. Decades of experimental and theoretical considerations predicted that AChE hydrolyzes ACh through a catalytic triad and depicted the active site as being comprised of two subsites: an anionic substrate binding site and an esteratic site (Rosenberry, 1975; Quinn, 1987). Catalysis was assumed to take place via formation of an acyl-enzyme intermediate followed by hydrolysis of the acyl-group to regenerate the enzyme.

However, debate regarding the components of the triad, the nature of the "anionic" subsite, and the mechanisms explaining the high turnover rate persisted until the three-dimensional structure of the enzyme was solved.

The crystallization and X-ray analysis of *Torpedo* AChE marked a milestone in AChE research (Sussman et al., 1991). First, the X-ray data revealed the components of the triad: serine 198, histidine 438, and glutamate 327. Second, the X-ray data placed the active site at the bottom of a 20-Å deep gorge lined with aromatic, not anionic, residues. The unusual spatial orientation of the catalytic triad and its inclusion of glutamate rather than aspartate distinguished AChE from the family of serine proteases and identified it with the superfamily of carboxyhydrolases, which includes a lipase from *Geotrichum candidum* (Schrag et al., 1991). The availability of cloned ACHE and BCHE DNAs combined with the resolution of the three-dimensional structure of *Torpedo* AChE provided the impetus for dozens of site-directed mutagenesis studies, which together are filling in important details on the ligand-binding interactions and catalytic mechanisms of the ChEs (Soreq et al., 1992; Taylor and Radic, 1994; Schwarz et al., 1995b). Utilizing the X-ray crystallography data and a computer simulation of ACh docking onto the enzyme, the gorge-lining residues were proposed to generate an electric dipole sufficient to account for the strong attraction of substrate to the active site (Ripoll et al., 1993). Based on a similar study, an "open back door model" was proposed to explain the inexplicably rapid diffusion of the catalytic byproducts away from the catalytic center (Gilson et al., 1994). However, site-directed mutagenesis of human AChE indicated that the electric dipole probably does not contribute much to the catalytic efficiency of the enzyme (Shafferman et al., 1994).

Acetylcholinesterase Molecular Polymorphism

AChE presents a complex panoply of molecular forms that can be distinguished through differential extraction procedures, sucrose-density centrifugation, and nondenaturing gel electrophoresis (Massoulie and Bon, 1982; Bon et al., 1991) (Fig. 13). AChE polymorphism is established by both posttranscriptional and posttranslational events that are tissue-specific (Rakonczay and Brimijoin, 1988) and developmentally regulated (Zakut et al., 1985).

AChE-T catalytic subunits may exist as soluble, globular, monomers (G1), dimers (G2), or tetramers (G4) (Fig. 13 and Table 5). AChE-T monomers and dimers of the nonamphiphilic ($G1^{na}$, $G2^{na}$) and type II amphiphilic ($G1^{a}$, $G2^{a}$) forms have been described (Bon et al., 1991). Interaction of the soluble, type II, amphiphilic forms with detergents is detectable in nondenaturing gel electrophoresis or density centrifugation, and may be an intrinsic character of the E6-encoded C-terminal peptide (Duval et al., 1992b; Massoulie et al., 1993). AChE-T tetramers may exist as nonamphiphilic homo-oligomers ($G4^{na}$), or become associated with noncatalytic structural subunits to produce a brain-specific amphiphilic form ($G4^{a}$) or a series of asymmetric "tailed" forms comprised of one (A4), two (A8), or three (A12) tetrameric units (Fig. 13 and Table 5). AChE-H subunits give rise to a GPI-linked, type I amphiphilic dimer ($G2^{a}$) found on vertebrate erythrocyte membranes and *Torpedo* electric organ (Fig. 13). The human readthrough AChE mRNA encodes soluble hydrophobic AChE-R monomers when expressed using heterologous expression protocols (Karpel et al., 1994,1995; Seidman et al., 1995), but no protein has yet been associated with the human readthrough AChE mRNA in vivo.

The principal brain and muscle forms of AChE are derivatives of AChE-T (Massoulie et al., 1993; *see also* Roberts et al., 1991). In muscle, a significant fraction of AChE is in the A12 configuration, comprised of a triplet of AChE-T tetramers covalently linked to a 50-nm-long, triple-helical, collagen-like, tail subunit. A12, as well as the other asymmetric AChE species, is concentrated at NMJs (Hall, 1973; Dreyfus et al., 1983) and is characterized by revers-

Table 5
ACHE Molecular Forms

Molecular form	S value[a]	Noncatalytic subunits	Hydrodynamic properties	Catalytic subunit/3' exon
A12	16–20 S	Collagenic tail (Q subunit)	Nonamphiphilic, aggregates in low ionic strength, sensitive to collagenase	AChE-T/E6
A8	14–15 S	Collagenic tail (Q subunit)		
A4	8–9 S	Collagenic tail (Q subunit)		
$G4^{na}$	10–11 S	None	Low salt soluble, nonamphiphilic	AChE-T/E6
$G4^{a}$	9–10 S[b]	20-kDa (P) subunit	Detergent soluble, amphiphilic	AChE-T/E6
$G2^{a}$ type I	6 S[b]	GPI-anchor	Partially salt soluble, amphiphilic	AChE-H/E5
$G2^{a}$ type II	6 S[c]	—	Easily solubilized without detergent, amphiphilic	AChE-T/E6
$G1^{a}$	4 S[c]	—	Amphiphilic, but easily solubilized without detergent	AChE-T/E6
$G1^{na}$	3–5 S[d]	None	Low salt soluble, nonamphiphilic[d]	AChE-R/I4[d]

[a]In presence of Triton X-100.
[b]Aggregates in absence of detergent.
[c]Shifts 1 S in absence of detergent but does not form aggregates.
[d]Characterized in heterologous expression systems only.
Adapted from Massoulie et al. (1993).

ible aggregation in low ionic strength, a high Stokes radius, and sensitivity to collagenase. As viewed by electron microscopy, the asymmetric forms give a balloon-like appearance (Dudai et al., 1973).

A12 is solubilized by high ionic strength (1*M* NaCl) in the presence of ethylenediaminetetraacetic acid (EDTA). These conditions reflect the strong polyanionic interactions that associate it with the basal lamina. Heparan sulfate proteoglycan has been strongly implicated with the anchorage of asymmetric AChE within the ECM (Brandan et al., 1985). However, the involvement of glycosaminoglycans of the chondroitin/dermatan sulfate family in ECM-association of asymmetric AChE forms cannot be excluded (Massoulie et al., 1993). The requirement for a calcium chelator to completely solubilize asymmetric AChE from muscle may reflect a maturation in the interaction of some forms (A^{ii}) with the basal lamina (Massoulie et al., 1993). Noteworthy is the observation that globular AChE forms also bind heparan and other glycosaminoglycans and could therefore potentially associate with ECM components (Ramirez et al., 1990).

A cDNA clone encoding the collagenic tail subunit (Q) has been isolated from *Torpedo* electric organ (Krejci et al., 1991). The predicted primary amino acid sequence contains a collagen-like core bounded by a proline-rich N-terminal domain and proline/cysteine-rich C-terminal domain. Three intertwined Q polypeptides are apparently disulfide-linked by cysteine residues on either side of the core to form the triple-helical structure. Coexpression of the collagenic subunit with the catalytic AChE-T subunit in COS cells gave rise, in addition to all the globular forms, to asymmetric A12 AChE, suggesting that all levels of assembly occur spontaneously. In that case, the generation of asymmetric AChE forms in nerves and muscle could be contingent solely on expression of the tail subunit. Thus, heterogeneity among mRNAs encoding the collagenic subunit was taken to imply that multileveled regulation of this gene-product could mediate the appearance and character of asymmetric AChEs (Krejci et al., 1991). Mutagenesis studies subsequently demonstrated that the covalent interactions between the catalytic and noncatalytic subunits are dependent on the C-terminal peptide of AChE-T and the N-terminal domain of Q (Duval et al., 1992a).

The predominant form of AChE in mammalian brain (approx 80%) is composed of a single G4 tetramer of AChE-T catalytic subunits covalently linked to a 20-kDa hydrophobic proteolipidic structural subunit ($G4^a$; Gennari et al., 1987; Inestrosa et al., 1987; Fuentes et al., 1988; Boschetti et al., 1994) (Fig. 13) through which the enzyme is anchored to membranes. This form is enriched in the caudate nucleus and is extractable with detergent. The remaining 20% of AChE activity in mammalian brain is low salt soluble (LSS)—40% G1 and 60% G4 (Liao et al., 1994). Several MAbs selective for detergent soluble (DS) brain AChE have been prepared (Liao et al., 1992,1993a,b).

Type I amphiphilic, dimeric, erythrocyte AChE, as well as amphiphilic dimers from *Torpedo* electric organ, *Xenopus* muscle, and *Drosophila* nervous system, carries an ethanolamine GPI anchor, which mediates the enzyme's association with the cell surface (reviewed by Silman and Futerman, 1987) (Fig. 13). Most of the GPI-linked AChEs are sensitive to phosphatidylinositol-specific phospholipase C (PIPLC) or phospholipase D (PLD). Enzymatic cleavage of the anchor with phospholipase releases AChE as a soluble, nonamphiphilic dimer. However, human erythrocyte AChE, in which the inositol ring is acylated, is resistant to PIPLC and requires treatment with alkaline hydroxylamine to yield a PIPLC-sensitive form or to render a PLD-treated preparation completely nonamphiphilic (Toutant et al., 1989). The specific structure of the GPI-anchor appears to be dependent on the host cell (Massoulie et al., 1993).

Human erythrocyte AChE can only be derived from AChE mRNA-bearing exons E2–E4 spliced to alternative exon E5. As predicted from the ACHE gene sequence (Karpel et al., 1994), the primary translation product of this RNA should be 586 amino acids. The 29 C-terminal amino acids constitute a cleavable hydrophobic peptide characteristic of precursors of GPI-linked proteins (Ben Aziz-Aloya, 1994). Soon after translation, the C-terminal peptide should be removed and replaced by covalent attachment of the GPI anchor to the C-terminal amino acid to generate the amphiphilic AChE-H of 557 (Ben Aziz-Aloya, 1994).

Heterologous Expression of Acetylcholinesterase

The first successful heterologous expression of catalytically active AChE was achieved in microinjected oocytes of *Xenopus laevis* from polyA^+ mRNA prepared from *Torpedo* electric organ (Soreq et al., 1982). Subsequently, microinjection of size fractionated polyA^+ RNA from primary tumors yielded evidence for heterogeneity of AChE mRNA (Soreq et al., 1984). Following cloning of a cDNA encoding human serum BuChE (Prody et al., 1987), studies on the tissue-specific processing (Dreyfus et al., 1989; Soreq et al., 1989) and ligand-binding properties (Neville et al., 1990a,b) of human BuChE were initiated. Finally, using oocyte microinjection to screen for AChE-mRNA-enriched human tissues, human AChE was cloned and expressed in oocytes from in vitro-transcribed mRNA (Soreq et al., 1990).

The ligand-binding and catalytic properties of oocyte-produced ChEs appear absolutely faithful to those of the native enzymes (Soreq et al., 1989; Seidman et al., 1994; Loewenstein-Lichtenstein et al., 1995; Schwarz et al., 1995a,b). Therefore, the availability and oocyte-expressibility of cloned human AChE and BuChE provided the opportunity to examine and compare the biochemical properties of a variety of natural and site-directed mutants of both ChE genes (Neville et al., 1992; Gnatt et al., 1994; Schwarz et al., 1995a), including an AChE-BuChE chimera (Loewenstein et al., 1993). Microinjected *Xenopus* oocytes have also been used to express catalytically active *Drosophila* AChE (Fournier et al., 1992b; Mutero and Fournier, 1992). However, in that case, the C-terminal peptide was incompletely processed, yielding a hydrophobic but non-GPI-linked enzyme form (Fournier et al., 1992b).

Plasmid DNA carrying ACHE coding sequences downstream of either the CMV promoter-enhancer, the Rous Sarcoma Virus long terminal repeat, or SV40 promoter have been transiently expressed in human embryonic kidney 293 cells (Velan et al., 1991a,b; Kronman et al., 1992) and COS cells (Gibney and Taylor, 1990; Duval et al., 1992a,b; Legay et al., 1993b). Heterologous expression in transfected mammalian cells has been employed in numerous structure-function studies (reviewed by Massoulie et al., 1993; Taylor and Radic, 1994) and studies of the biosynthetic and secretory pathways of AChE (Velan et al., 1991a,1993; Shafferman et al., 1992a,b; Kerem et al., 1993). It should be noted, however, that <5% of the immunoreactive protein produced from *Torpedo* ACHEDNA in COS cells at 27°C was catalytically active; at 37°C, AChE activity was almost undetectable (Duval et al., 1992b). These observations were understood to imply thermal restrictions on correct folding of the *Torpedo* enzyme and could have implications for any cross-species heterologous expression of AChE.

Expression of the AChE-T subunits from *Torpedo* (Duval et al., 1992b) and rat (Legay et al., 1993a) in transiently transfected COS cells gave rise to all the globular AChE forms. In COS cells, approx 50% of AChE activity was secreted into the culture medium; of the remaining activity, 40–50% was associated with the external face of the cell membrane (Duval et al., 1992b). Cotransfections of AChE-T-encoding DNAs from both *Torpedo* and rat with DNA encoding the *Torpedo* collagenic tail subunit (Q) gave rise to the A12 asymmetric form characteristic of NMJ AChE, demonstrating evolutionarily conserved, spontaneous association of the catalytic and noncatalytic AChE subunits (Duval et al., 1992a,b; Legay et al., 1993b). When cDNAs encoding AChE-H from several species were expressed in COS cells, membrane-associated, PIPLC-sensitive, amphiphilic AChE forms were obtained, indicating the production of GPI-linked forms. Together, the cumulative results of heterologous expression studies provide strong experimental support to the contention that alternative usage of 3' exons in the ACHE gene combined with the use of variable noncatalytic subunits accounts for all the known molecular forms of AChE.

Expression of catalytically active AChE has also been achieved using baculovirus transfection of *lepidoptera* cells (Radic et al., 1992) and transformation of *Escherichia coli* (Fischer et al., 1993). Recovery of active AChE from *E. coli* required replacement of GC residues in the ACHE coding sequence with conserved AT substitutions and the application of *post-facto,* in vitro folding of catalytically inactive polypeptides. Devoid of the C-terminal cysteine required for disulfide linkage of AChE monomers, the bacterial enzyme did not display oligomeric assembly. Moreover, bacterially produced AChE was necessarily bereft of glycosylation. This latter point corroborates the findings of Dreyfus et al. (1989) and Velan et al. (1993), indicating that glycosylation of human ChEs is a prerequisite for secretion, but not for catalytic activity.

To explore the tissue-specific expression of AChE, the role of AChE during development, and the involvement of AChE in various neuro- and myopathologies, in vivo models are required. To date, three animal models have been reported: *Drosophila* (Hall and Spierer, 1986; Fournier et al., 1992a), *Xenopus* (Ben Aziz-Aloya et al., 1993a; Seidman et al., 1994,1995; Shapira et al., 1994), and a single line of transgenic mice (Beeri et al., 1995). Overexpression of a truncated soluble form of *Drosophila* AChE in the body fluids of transfected flies provided protection against several anti-AChE insecticides (Fournier et al., 1992a). In transgenic mice, the human AChE-coding sequence, under control of the proximal human ACHE promoter, was expressed in CNS neurons only, as determined by RT-PCR and *in situ* hybridization. Overexpressed AChE conferred protection from the hypothermic effects of organophosphorous AChE inhibitors and promoted an age-dependent impairment in spatial learning and memory resembling that observed in AD (Beeri et al., 1995). Although human AChE RNA was not observed in muscle, human AChE protein was nevertheless detectable in muscle extracts. Enhanced cytochemical staining for catalytically active enzyme in NMJs could therefore be attributed to motoneuron-produced human AChE. NMJs from diaphragms of the transgenic mice displayed enlarged overall structure and greatly exaggerated secondary folding of the postsynaptic membrane, which was correlated with weakness and fatigue characteristic of myopathology (Andres et al., in preparation).

In microinjected *Xenopus* embryos, AChE-encoding DNA downstream of the CMV promoter produced transiently high levels of overexpressed enzyme, which accumulated in muscles and NMJs. It is this last system to which the remainder of this monograph will be devoted.

Chapter 2

Experimental Methodologies

This chapter is meant to serve two functions. First, we present the methods used to produce the experimental results presented in the following chapters. These are designed to assist the reader in understanding the accompanying figures and to define the specific experimental conditions used to produce them. Among them are methods relating specifically to the use of *Xenopus* as well as those representing more mundane laboratory procedures such as gel electrophoresis and sucrose gradient centrifugation, together with protocols whose utilities are more restricted to those of us working on ChEs. We also present several appendices (*see* pp. 147–165) whose purpose is to provide detailed information and protocols relating to *Xenopus* husbandry and microinjections. Naturally, these appendices are based primarily on our own laboratory experience, but we have tried to incorporate the experiences of others working in the field, where our own experience is limited.

Reagents, Buffers, and Solutions

1. Modified Barth's medium (MBM): 88 m*M* NaCl, 1 m*M* KCl, 0.33 m*M* $Ca(NO_3)_2$, 0.82 m*M* $MgSO_4$, 2.4 m*M* $NaHCO_3$, 0.4 m*M* $CaCl_2$, 20 m*M* HEPES, pH 7.4; (*see* Appendix III for recipe).
2. Mark's modified ringer (MMR): 1X MMR—100 m*M* NaCl, 2 m*M* KCl, 1 m*M* $MgSO_4$, 2 m*M* $CaCl_2$, 5 m*M* HEPES, pH 7.4, 0.1 m*M* EDTA (*see* Appendix III for recipe).
3. Phosphate-buffered saline (PBS): 0.01*M* sodium phosphate, pH 7.4, 144 m*M* NaCl.
4. TTNE: 0.01*M* Tris, 1.0*M* NaCl, 1% Triton X-100, 1 m*M* EGTA, pH 7.4.
5. Low salt (LS) buffer: 0.02*M* Tris-HCl, pH 7.5, 0.01*M* $MgCl_2$, 0.05*M* NaCl.

6. Low salt detergent (LSD) buffer: 0.01*M* sodium phosphate, pH 7.4, 1% Triton X-100.
7. High salt (HS) buffer: 0.01*M* sodium phosphate, pH 7.4, 1.0*M* NaCl, 1 m*M* EGTA.
8. MEMFA: 0.1*M* MOPS, 2 m*M* EGTA, 1 m*M* $MgSO_4$, 3.7% formaldehyde, pH 7.4.
9. PBT: PBS + 0.1% Triton X-100.
10. Diaminobenzoic acid (DAB): **Suspected carcinogen!** Prepare 10 mg/mL stock in double-distilled water (ddH_2O), filter to remove insoluble material, and store frozen in 1-mL aliquots.
11. Murray's clearing solution (benzyl alcohol:benzyl benzoate 1:2 [BABB]): **Caustic!** Wear gloves when handling and avoid immersing microscope lens in clearing solution.

Microinjections

Vectors

In Vitro-Transcribed RNA

For preparation of in vitro-transcribed AChE mRNA, the AChE coding sequence (Soreq et al., 1990) was subcloned into *pGEM* transcription vector downstream of the SP6 RNA polymerase binding site (Promega, Madison, WI). In vitro transcription was performed using SP6 RNA polymerase according to the manufacturer's instructions.

Fig. 15. *(opposite page)* DNA expression vectors. Several DNA constructs employed in these studies and described in detail in this chapter are presented. **(A)** AChE-encoding constructs: ACHE-I4/E5 was constructed from ACHE-E6 and a genomic ACHE clone. ACHE-E6 encodes brain and muscle AChE (AChE-T); ACHE-I4/E5 potentially encodes erythrocyte AChE (AChE-H) or readthrough AChE (AChE-R) through alternative usage of 3' exons. The human ACHE promoter-reporter plasmid used to define the basal promoter region is depicted in Fig. 22. **(B)** AChR-encoding plasmids: cDNA clones encoding the individual subunits of AChR were removed from vector pSP65 and placed in pcDNA I using the restriction sites indicated in the figure. The heavy line in the figure indicates the portion of the clone encoding the mature subunit; the hatched box indicates the multiple cloning site of pcDNA I, and the SV40 promoter is shown as the box to the left of the *Hin*dIII cloning site. Reprinted from Shapira et al. (1994) with permission from National Academy of Sciences.

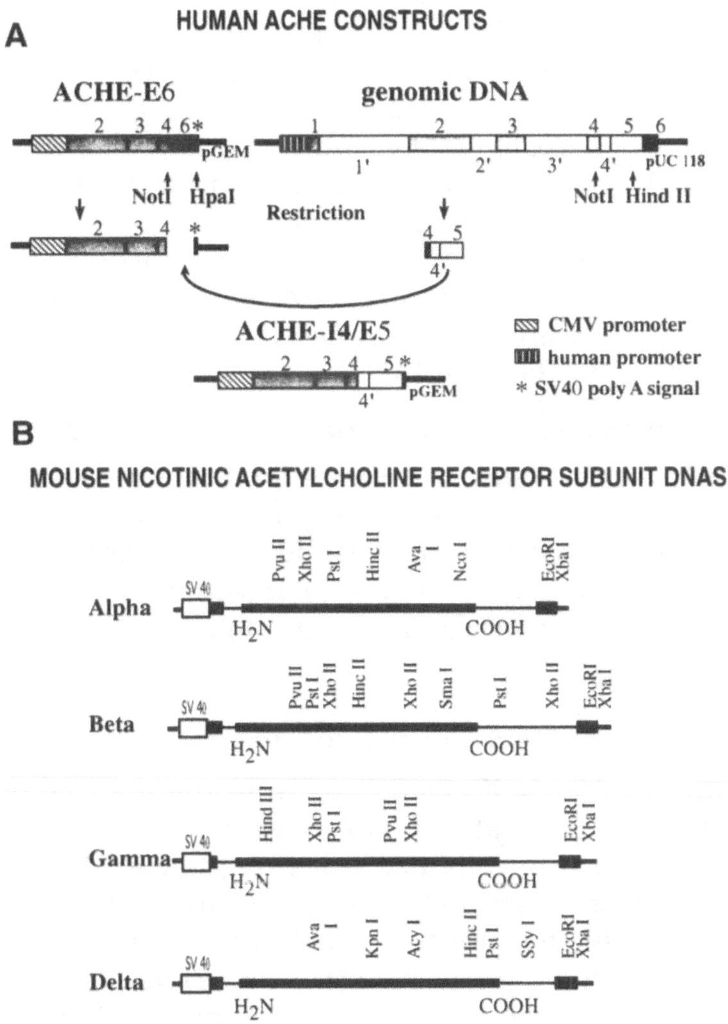

Fig. 15.

DNA Expression Plasmids—Acetylcholinesterase

The plasmid referred to here as ACHE-E6 and employed to express the major brain and muscle form of AChE (AChE-T) has been described in detail (CMVACHE; Velan et al., 1991b). This plasmid contains the AChE coding exons E2–E4 and E6 (Soreq et al., 1990) (Fig. 15) downstream of the CMV promoter and followed by the SV40 polyadenylation site. ACHE-E6 was used to construct ACHE-I4/E5 by exchanging the cDNA restriction fragment *Not* I-*Hpa* I with the genomic fragment *Not* I-*Hind*II (Fig. 15). ACHE-I4/E5 carries exons E2–E4, intron I4, and exon E5 and potentially encodes both AChE-H and/or AChE-R. The human acetylcholinesterase promoter-reporter construct (HpACHE) carries the AChE coding sequence downstream of the human ACHE gene promoter and was derived from genomic clones and cDNA clones as described (Ben Aziz-Aloya et al., 1993a) (Fig. 14).

DNA Expression Plasmids—Acetylcholine Receptor

cDNA clones encoding the 4 subunits of mouse muscle nAChR were placed under control of the SV40 promoter (Fig. 15). Prior to injection, plasmids were mixed in a ratio of 2:1:1:1 (α, β, γ, δ, respectively). Injections were approx 1 ng/embryo, total DNA. AChR plasmids were a gift from J. Patrick.

Xenopus *Oocyte Microinjections*

Mature female *Xenopus* frogs were anesthetized on ice (40–60 min). Several ovarian lobes containing 500–600 oocytes were surgically excised through a small incision in the lower abdomen and placed in 1X MBM. Single sutures through both the muscle and skin were performed to close the wound. Operated frogs could be reused after complete healing of the incision, about 3 mo after the operation, or by making a second incision on the opposite side of the abdomen following 4–6 wk rest. All subsequent oocyte manipulations, including microinjections, were performed in 1X MBM.

Oocytes were manually teased from the ovarian tissue using a pair of fine-tipped forceps, periodically replacing turbid buffer with fresh MBM. Alternatively, oocytes may be released from the ovarian tissue using collagenase (*see* Appendix VII). Single, mature, stage 5–6 oocytes

(Dumont, 1972) were selected by eye, removed to a separate dish, and used for microinjection.

For microinjections, oocytes were placed into small wells in a plexiglass dish. Fifty to one hundred nanoliters of RNA (0.1 mg/mL) or 5–10 nL DNA (1 mg/mL) in sterile ddH_2O were injected into each oocyte. Following injection, oocytes were incubated in a humidified incubator at 17°C. The incubation medium was changed once at the end of the day and then once every 24 h. Dead oocytes (as evidenced by loss of pigment) were removed whenever noticed. Microinjected oocytes could be maintained for up to 5 d in reasonably good health. For some DNA injections, oocytes were centrifuged (600*g*) for 10 min to bring the nucleus to the surface where it was visible as a small white protrusion. However, this procedure resulted in the death of many oocytes and did not always work. Therefore, in most experiments, oocytes were injected "blind" deep into the animal pole. In that case, variability in expression between oocytes was high (10–20-fold), but homogenates prepared from pools of 5–10 oocytes gave satisfactory results.

Xenopus *Embryo Microinjections (see Appendix V for Detailed Protocol)*

Mature females were injected with HCG (500–800 U; Sigma, St. Louis, MI) (Fig. 16A) 24 h prior to the experiment to induce spawning. Mature fertilizable eggs were released into a dry petri dish by applying gentle pressure to the lower back and abdomen while holding the legs apart (Fig. 16B). Males were sacrificed by anesthetic overdose (MS222; Sandoz, Basel, Switzerland; 200 mg) and the testes were excised. In vitro fertilization was accomplished by dragging a small piece of macerated testis over the eggs and flooding the dish with 0.1X MMR after 90 s.

Fertilized eggs were dejellied with 2% cysteine (pH 7.5; 2 × 3 min) (*see* Appendix III), rinsed five to six times with 0.1X MMR, and allowed to rest undisturbed. At approx

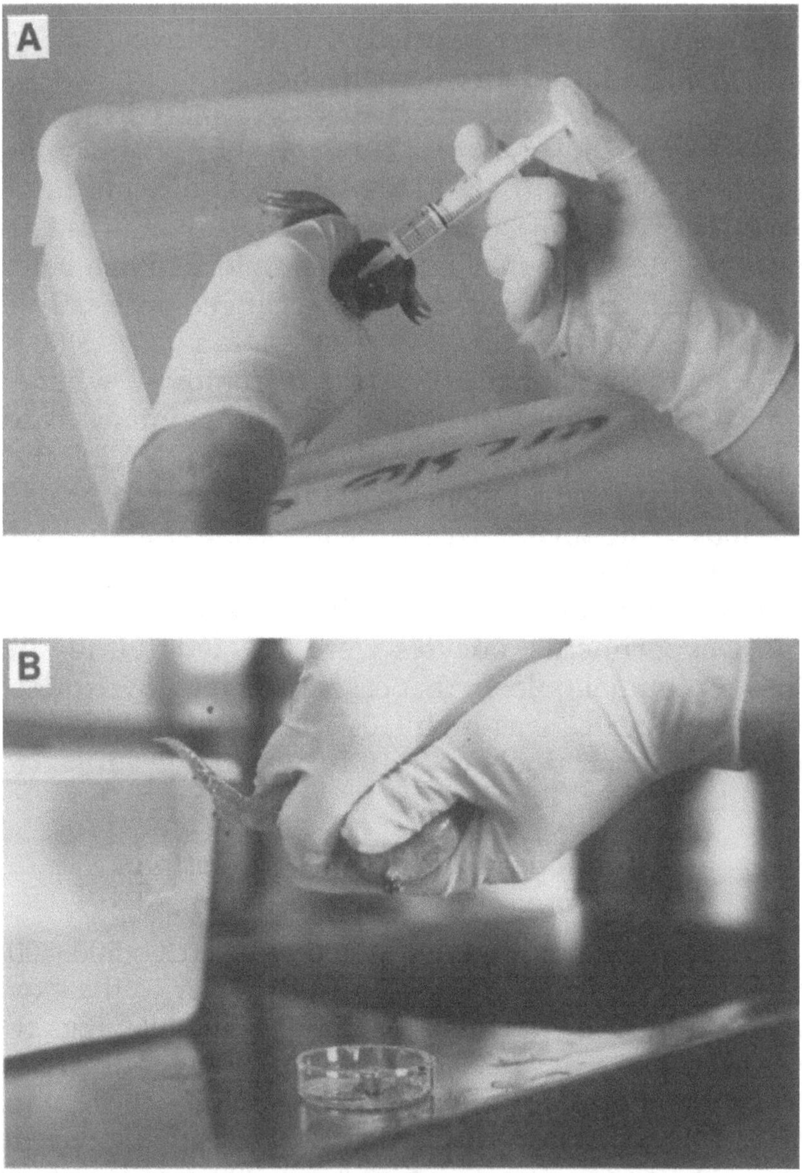

Fig. 16. Preparing eggs for in vitro fertilization. Approximately 16–18 h prior to in vitro fertilization, several female frogs are injected with HCG into the dorsal lymph sack **(A)**. The hormone treatment promotes maturation of oocytes into fertilizable eggs and stimulates spawning. Eggs are collected by firmly holding the frog with both hands and applying gentle pressure to the back and abdomen **(B)**.

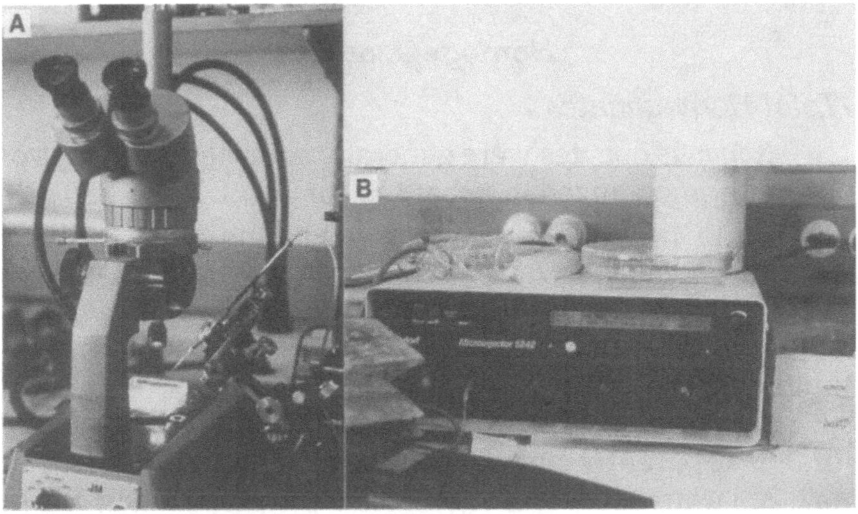

Fig. 17. Microinjection equipment. Microinjection of *Xenopus* oocytes and embryos is accomplished under a dissecting binocular accompanied by an X-Y-Z micromanipulator and illuminated by an optic-fiber cold-light source **(A)**. Use of a high-pressure microinjection unit for dispensing injection volumes allows for utilization of a fine-tipped microinjection needle to minimize injection-related damage to embryos **(B)**.

70 min postfertilization, embryos were transferred to 0.3X MMR containing 5% ficoll-400 and injected within the first 2 cleavage cycles using a high pressure microinjector (Fig. 17). To slow the cleavage rate, embryos could be cooled to as low as 15°C for several hours without apparent negative aftereffects. For histochemical studies, only embryos injected prior to completion of the first cleavage were used. Several hours after microinjection, embryos were transferred by successive dilutions into 0.3X MMR without ficoll and cultured overnight at 17–21°C. One-day-old embryos were sorted and healthy gastrulae were transferred to either 0.1X MMR or aged tap water. From d 4 or d 5 postfertilization, frozen spinach was added as a food source, and the water was gently aerated.

Biochemical Analyses

Homogenizations

Total Homogenates

Xenopus oocytes were generally collected in groups of 10 and homogenized in either LSD or TTNE buffer (30 µL/ oocyte) in an Eppendorf tube using a hand-held, battery-powered homogenizer (Kontes, Vineland, NJ). Homogenates were microfuged for 10–20 min in the cold and the clear supernatant was removed to a clean tube with a drawn-out Pasteur pipet.

Xenopus embryos were generally harvested in groups of 3–5 apparently normal individuals and stored frozen (–20°C) until used. Homogenates were prepared in TTNE (100–150 µL/embryo) and assayed for enzymatic activity (1/10 embryo/assay) following a 10–15-min centrifugation in the cold. Through the tailbud stage, embryos could be homogenized by repetitive pipeting through a microtip. From d 3 postfertilization, homogenization was performed in a glass-Teflon homogenizer to ensure complete disruption of the tissues.

Subcellular Fractionation

For subcellular fractionations, groups of 3 embryos were homogenized in LS buffer (100 µL/embryo) in a glass-Teflon homogenizer, and centrifuged at 400,000*g* for 10 min in a Beckman TL100 tabletop ultracentrifuge. The supernatant was collected and considered the low salt soluble fraction. The pellet was resuspended in LSD buffer, incubated on ice for 1 h, and centrifuged as above for 5 min to generate the detergent soluble fraction. The remaining pellet was resuspended in HS buffer to release the high salt soluble (HSS) AChE fraction. In some cases, enzyme antigen immunoassay (EAIA) (*see* Enzyme Antigen Immunoassay section in this chapter) was employed to distinguish between recombinant human acetylcholinesterase (rHAChE) and *Xenopus* AChE in the fractions.

Acetylcholinesterase Activity Assays

AChE activity was evaluated in all experiments using a modified version of the colorimetric assay (devised by Ellman et al., 1961) adapted to a 96-well microtiter plate. Ten to twenty microliters of enzyme were assayed in 200 µL (final volume) 0.2*M* phosphate buffer, pH 7.4, 0.5 m*M* 5-5' dithiobis-nitrobenzoic acid (DTNB) containing 1 m*M* acetylthiocholine iodide (Sigma) substrate. For inhibitor studies, samples were preincubated with specific inhibitors for 30–40 min prior to the addition of substrate. For selective inhibition of rHAChE in embryo homogenates, a 30-min preincubation with echothiophate ($3.3 \times 10^{-7} M$) was employed. A_{405} was monitored at 3–5-min intervals for 20–30 min using a VMAX microtiterplate reader (Molecular Devices Corporation [MDC], Palo Alto, CA), and the rate of hydrolysis was calculated by regression analysis using a dedicated software package (SoftMax, MDC). All assays were performed at room temperature. Raw data was translated to activity units (nanomoles substrate hydrolyzed/embryo/h) using the extinction coefficient for DTNB = $13{,}600 M^{-1}cm^{-1}$ and a path length of 0.5 cm.

Sucrose Gradient Ultracentrifugation

Freshly prepared HS/detergent extracts from 1–2 embryos or 5–10 oocytes were applied to 12 mL 5–20% linear sucrose density gradients and centrifuged overnight at 4°C. Fractions were collected into 96-well microtiter plates and assayed for total AChE activity, essentially as described in Acetylcholinesterase Activity Assays section in this chapter, but using an endpoint reading after one to several hours of incubation at room temperature. When necessary to distinguish between rHAChE and endogenous *Xenopus* AChE in the gradient fractions an EAIA was employed (*see* the next section). Briefly, 100-µL aliquots were transferred to a Maxisorp immunoplate (Nunc, Denmark) coated with a MAb (101-1) recognizing human but not *Xenopus* AChE, and diluted 1:1 with ddH_2O. Following overnight incubation, the plates were washed three times with PBS containing 0.05% Tween-20 and each well was assayed for catalytically active AChE.

Enzyme Antigen Immunoassay

EAIA was exploited in some experiments to permit selective analysis of rHAChE when significant levels of endogenous *Xenopus* AChE were present in the homogenates. Briefly, wells of a Maxisorp immunoplate (Nunc) were incubated overnight at 4°C with 100 µL (5 µg/mL) of a species-specific antihuman AChE MAb (101-1) in 0.1*M* carbonate buffer (pH 9.6). The plate was rinsed and incubated for at least 1 h with 1% bovine serum albumin (BSA) to block nonspecific binding sites. To each well, up to 50 µL homogenate were added and incubated overnight in the cold. The plates were washed three times with PBS containing 0.05% Tween-20; each well was assayed for catalytically active AChE.

Polyacrylamide Gel Electrophoresis

Sodium Dodecyl Sulfate-Polyacrylamide Gel Electrophoresis/Immunoblot

rHAChE was purified from approx 180 1-d-old DNA-injected embryos by affinity chromatography using a modified procedure for the purification of native human AChE (Gennari and Brodbeck, 1985). AChE from embryos homogenized in LSD buffer was bound to Sepharose beads carrying N-(1-amino-hexyl) 3-dimethylethylaminobenzoic amide (gift from U. Brodbeck, Switzerland) by shaking overnight at room temperature. Elution was with 0.02*M* edrophonium chloride (Tensilon, Hoffmann-LaRoche, Switzerland). Embryonic *Xenopus* AChE was similarly purified from 1-wk-old tadpoles, but had to be eluted by boiling in 0.1% SDS. Denaturing sodium dodecyl sulfate-polyacrylamide gel electrophoresis (SDS-PAGE) and blotting were essentially as described (Liao et al., 1992) using a pool of MAbs (132-1, 132-2, 132-3; 6 μg/mL each) raised against denatured human brain AChE (Liao et al., 1993a,b).

Nondenaturing Gel Electrophoresis

Nondenaturing gel electrophoresis was performed in vertical 6% polyacrylamide gels using a 3% stacking gel. Gel and running buffers were identical to those employed for standard Laemelli SDS-PAGE (Sambrook et al., 1989), except that SDS was omitted from all solutions; 1% Triton X-100 was substituted for detergent where appropriate. Electrophoresis was performed at 120 V for 2–4 h in the cold. The gel was rinsed three times with ddH_2O, and stained for catalytically active AChE for several hours to overnight using the histochemical staining procedure of Karnovsky and Roots (1964) (*see* the next section and Appendix VI).

Histochemical Analyses

Whole-Mount Cytochemical Staining

Two-day-old *Xenopus* embryos were fixed for 20 min in 4% paraformaldehyde (in 0.6X PBS), rinsed three times with PBS, and transferred to a clean glass vial (15 × 45 mm). Fixed embryos were incubated in staining solution (0.67 m*M* acetylthiocholine, 5 m*M* sodium citrate, 3 m*M* cupric sulfate, 0.5 m*M* potassium ferricyanide in 0.1 *M* acetate buffer, pH 5.9) [Karnovsky and Roots, 1964; *see* Appendix VI]) overnight at room temperature with gentle rotation, rinsed several times with PBS, and postfixed with 2.5% glutaraldehyde in PBS for 1 h. Embryos were then dehydrated by successive transfers through 30–50–70–100% methanol, mounted in Murray's clearing solution (BABB 1:2) and viewed under low magnification with a Zeiss (Oberkochen, Germany) stereomicroscope. Clearing permitted visualization of internal structures and improved with time up to 18–24 h. For high magnification light microscopy, stained embryos were cut in half, transferred stepwise into 70% ethanol and then into 100% ethanol overnight using an automated dehydration system. Embryos were embedded in paraffin and cut with a microtome to generate 4–6-μm sections. Counterstaining was "modified three chromostain," including hematoxylin, eosin, and phosphomolybdic acid/light green stain.

Whole-Mount Immunocytochemical Staining

Two-day-old embryos were fixed for 1 h in MEMFA (0.1*M* MOPS, 2 m*M* EGTA, 1 m*M* $MgSO_4$, 3.7% formaldehyde, pH 7.4) at room temperature, washed 2 × 15 min in PBS, and 1 × 15 min in PBT. Nonspecific binding sites were blocked by preincubation for 1 h at room temperature in PBT containing 10% fetal calf serum (FCS; heat inactivated for 30 min at 56°C). Incubation with pooled MAbs raised against native bovine brain AChE (MAb 101-1, 101-2, 111-4; 3 µg each) was in PBT/FCS overnight at 4°C with gentle rotation. Primary antibodies were removed and embryos were washed with PBT 3 × 2 h and incubated overnight 4°C with horseradish peroxidase (HRP)-conjugated goat antimouse antibody (Fab fragment; Jackson Labs, West Grove, PA; 1:500) in PBT/FCS. Second antibody was removed and embryos were washed 4 × 2 h with PBT including one overnight wash at 4°C. Chromogenic staining was accomplished by preincubating embryos with DAB (0.5 mg/mL) in 0.5X PBT (1 mL/vial) for 5 min with vigorous shaking followed by addition of 1 µL 15% hydrogen peroxide. Staining reaction was carried out on ice for 2–4 h with shaking. To stop the reaction, embryos were transferred to PBS overnight at 4°C. Embryos were postfixed with 2.5% glutaraldehyde in 100 m*M* phosphate buffer for 1 h, after which they were dehydrated by successive passes through 30–50–70–100% methanol. Visualization was accomplished by clearing in BABB and photographs taken under low magnification (2.5–10× objective) with a Zeiss Axioplan microscope.

Electron Microscopy

Two- and three-day-old embryos were fixed for 20 min in 1% paraformaldehyde/2.5% glutaraldehyde/0.1*M* sodium phosphate (pH 7.4), and cut into three or four pieces. Fixation was continued for another 40 min, after which the embryos were washed three to four times with PBS. Tissue was washed an additional two to three times with 0.1*M* acetate buffer (pH 5.9) and then incubated with staining solution (1.7 m*M* acetylthiocholine, 5 m*M* sodium citrate, 3 m*M* cupric sulfate, 0.5 m*M* potassium ferricyanide in 0.1*M* acetate buffer (pH 5.9) (Karnovsky, 1964; *see* Appendix VI) for 20 min on ice without shaking. Embryos were rinsed several times, transferred to 1% osmium tetroxide for 1 h, dehydrated in ethanol, and embedded in Epon. Thin sections close to the exposed surface were prepared (700 Å) and counterstained with uranyl acetate and lead citrate. Computer-assisted morphometric analyses were performed on electron micrographs using the Sigma Scan software package (Jandel Scientific, Berkeley, CA).

RT-PCR Procedure and Primers

Total RNA was extracted from *Xenopus* embryos 1 d after injection by the RNAsol-B method (CINNA/BIOTECX), according to the protocol supplied by the manufacturer, and was treated with DNase (Promega) for 20 min at 37°C, 20 U/sample in 40 m*M* Tris-HCl/10 m*M* NaCl/6 m*M* $MgCl_2$ in the presence of RNAsin (Boehringer Mannheim, Mannheim, Germany, 20 U/sample). DNase was heat inactivated by incubation at 90°C for 8 min as described (Ben Aziz-Aloya et al., 1993a). RT-PCR analyses were performed using a Perkin-Elmer/Cetus thermal controller. Reverse transcription was performed on 0.1–2 μg RNA using the Moloney murine leukemia virus (MMLV) reverse transcriptase (Gibco-BRL, Bethesda, MD). Amplification was for 1 min, 94°C, first cycle 3 min; 1 min, 65°C, 1 min, 72°C, last cycle 6 min, performed with human AChE specific primer pairs (Ben Aziz-Aloya et al., 1993a; Seidman et al., 1995). Amplification products (20%) were electrophoresed on a 2% agarose gel and UV-photographed (320 nm).

Chapter 3

Experimental Applications

Human Acetylcholinesterase as a Model Nervous System Protein

Xenopus Oocyte Microinjections

Human Acetylcholinesterase Expressed in mRNA-Injected Xenopus *Oocytes*

A DNA sequence encoding the brain and muscle form of human AChE (AChE-T) was constructed in our laboratory from cloned cDNA and genomic sequences, and tentatively identified by its homology to known ChEs (Soreq et al., 1990). This putative AChE-coding sequence, bearing the 3' alternative exon E6, was subcloned into the SP6 transcription vector from which in vitro-transcribed mRNA was prepared (*see* Experimental Methodologies in Chapter 2). Microinjected into mature *Xenopus laevis* oocytes, 5 ng in vitro-transcribed AChE mRNA directed the production of catalytically active recombinant human acetylcholinesterase (rHAChE), at levels of approx 10-fold above background oocyte levels (Table 6). Oocyte-produced rHAChE was sensitive to the AChE-specific reversible inhibitor 1,5-bis-(4-allyldimethylammoniumphenyl)-pentane-3-one dibromide (BW284C51/BW) and insensitive to inhibition by the irreversible BuChE-specific inhibitor tetraisopropylpyrophosphoramide (iso-OMPA), as expected for a mammalian AChE. At $10^{-4}M$ BW, endogenous oocyte AChE activity (Gundersen and Miledi, 1983) was only 50% inhibited, whereas rHAChE was essentially 100% inhibited at this concentration. A similar differential sensitivity of the amphibian

Table 6
Catalytically Active rHAChE Expressed
in Microinjected *Xenopus* Oocytes[a]

	Acetylthiocholine hydrolysis, nmol/h/oocyte	
	AChE mRNA-injected	Buffer-injected
No inhibitor	50.4 ± 5.0	5.2 ± 0.9
+ BW284C51	2.3 ± 0.2	2.5 ± 0.6
+ iso-OMPA	47.8 ± 0.1	4.6 ± 0.3

[a]In vitro-transcribed AChEmRNA encoding AChE-T was expressed in microinjected *Xenopus* oocytes as described in Experimental Methodologies in Chapter 2. Parallel groups were injected with modified Barth's medium as control. Activity in total homogenates was determined in the presence and absence of 10 μ*M* of specific AChE (BW284C51) or BuChE (iso-OMPA) inhibitors following a 40-min preincubation. Spontaneous hydrolysis of substrate was subtracted. Data represent nanomoles substrate hydrolyzed/h/oocyte ± standard evaluation of the mean (SEM) for three independent transcription reactions and microinjection experiments. Note residual endogenous, BW284C51-insensitive hydrolytic activity of approx 2.5 nmol/h/oocyte in both groups. Adapted from Seidman et al. (1994).

and mammalian enzymes to inhibition was noted for several other anti-ChE agents (Soreq et al., 1982) and later exploited to differentiate between the two in complex mixtures.

Heterologous Acetylcholinesterase Is Biochemically Indistinguishable from Native Human Acetylcholinesterase

rHAChE displayed substrate and inhibitor interactions characteristic of the native human enzyme. rHAChE exhibited inhibition by excess substrate (>3 mM) (Fig. 18A) and an apparent Km of 0.3 m*M* toward acetylthiocholine—essentially identical to that displayed by native human erythrocyte AChE (data not shown) and rHAChE expressed in human 293 cells (Velan et al., 1991b). Moreover, the inhibition profiles for rHAChE toward various selective AChE and BuChE inhibitors were indistinguishable from those for the native human enzyme (Fig. 18B). In sucrose density centrifugation, rHAChE sedimented primarily as

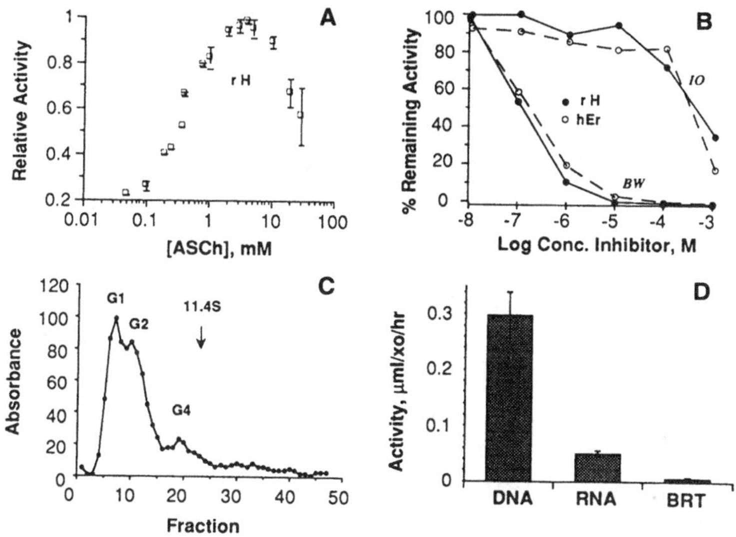

Fig. 18. Oocyte-expressed rHAChE. **(A)** Inhibition by excess substrate: Mature *Xenopus* oocytes were injected with 5 ng in vitro-transcribed AChE mRNA and incubated overnight at 17°C. Homogenates corresponding to 1/3 oocyte were assayed for AChE activity in the presence of various concentrations of acetylthiocholine substrate (average of three experiments ± SEM). **(B)** Sensitivity to selective inhibitors: Oocyte homogenates were preincubated for 30 min in assay buffer containing an AChE-specific inhibitor (BW) or a BuChE-specific inhibitor (IO) at the indicated concentrations and assayed for remaining activity following addition of 2 m*M* acetylthiocholine. AChE from human erythrocytes (hEr) served as control. **(C)** Oligomeric assembly: Homogenates from mRNA-injected oocytes were subjected to linear 5–20% sucrose density centrifugation. Note that in addition to the free monomer (3.2 S, G1), the oocyte appears to generate dimers (5.6 S, G2) and, to a lesser extent, tetramers (10.2 S, G4) of human AChE. Endogenous oocyte AChE activity is undetectable under these conditions. Arrow marks position of bovine liver catalase (11.4 S). **(D)** Expression of ACHE DNA in *Xenopus*: Oocytes were injected with 5 ng in vitro-transcribed ACHE mRNA or ACHE-E6 DNA and incubated for 1 (RNA) or 3 (DNA) d. Oocytes injected with incubation medium (BRT) or uninjected oocytes served as control. Activity is expressed as µmol substrate hydrolyzed/h/oocyte ± SEM for three independent microinjection experiments. Reprinted from Seidman et al. (1994) with permission from Raven Press.

monomers and dimers, although a discernible peak apparently representing globular tetrameric AChE was also observed (Fig. 18C).

Cytomegalovirus Promoter Directs Acetylcholinesterase Expression in DNA-Injected Xenopus *Oocytes*

When plasmid DNA (ACHE-E6) (Fig. 15) carrying the ACHE coding sequence downstream of the CMV promoter-enhancer element was microinjected into *Xenopus* oocytes, active AChE in yields 5–10-fold higher than that observed following RNA injections was obtained (Fig. 18D). The variability in expression between individual oocytes injected with DNA was much greater than that observed among RNA-injected oocytes, and may indicate inconsistencies in the success of nuclear injections. Nonetheless, by obviating the need for costly and time-consuming in vitro transcription reactions, DNA injections are the method of choice when practical.

Xenopus Embryo Microinjections

Transient Expression of Human Acetylcholinesterase in Microinjected Xenopus *Embryos*

To investigate the putative role of AChE in development, ACHE-E6 was expressed in *Xenopus* embryos. Microinjected into fertilized *Xenopus* eggs (Fig. 19),

---

Fig. 19. *(opposite page) Xenopus* embryo microinjection scheme. A schematic representation of a microinjection experiment depicting the principal developmental stages and analytical approaches used in this work is shown together with photographs displaying the normal overall development of unstained microinjected embryos (+) compared with control uninjected embryos (–) 3 d postfertilization. Note that microinjected embryos are slightly smaller than uninjected controls. Sketches adapted from Deuchar (1966).

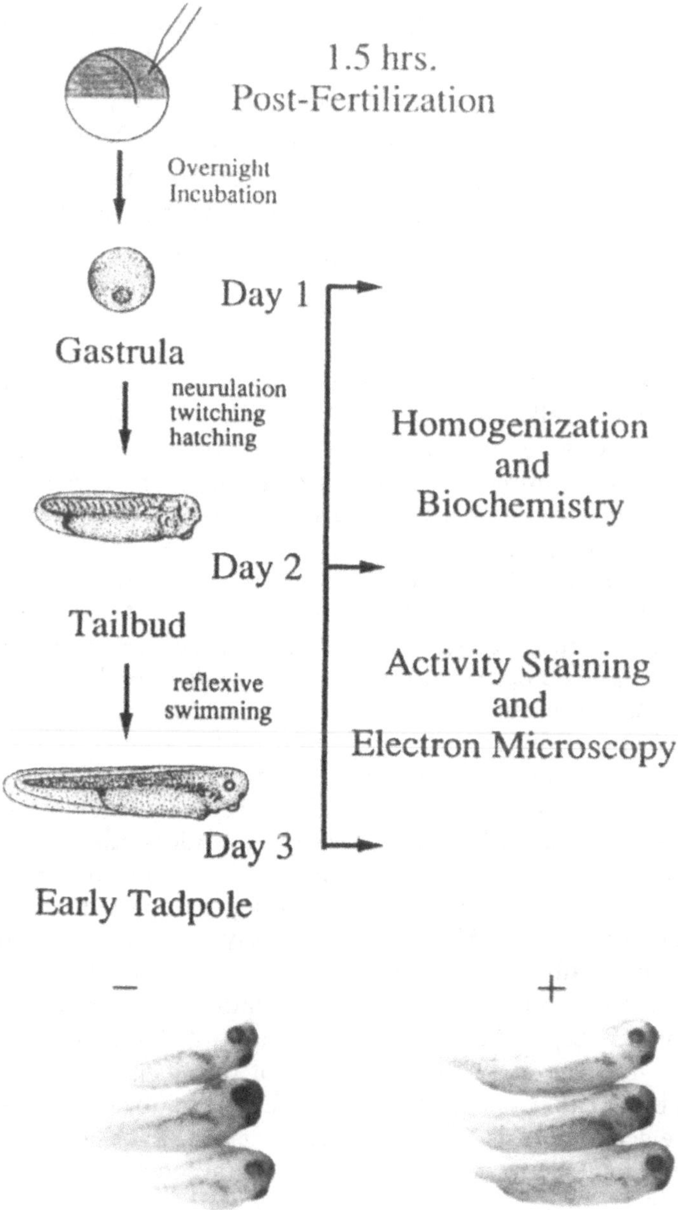
1.5 hrs.
Post-Fertilization
Overnight
Incubation
Day 1
Gastrula
neurulation
twitching
hatching
Homogenization
and
Biochemistry
Day 2
Tailbud
reflexive
swimming
Activity Staining
and
Electron Microscopy
Day 3
Early Tadpole
–
+

Fig. 19.

ACHE-E6 directed the biosynthesis of rHAChE at levels similar to those observed in DNA-injected oocytes. Following overnight incubation, at which time embryos had reached the late gastrula stage, endogenous AChE levels were negligible and rHAChE activity represented a 50–100-fold excess over normal (Fig. 20A). From d 2 postfertilization, detectable endogenous AChE activities increased steadily. Using the irreversible AChE inhibitor echothiophate to distinguish between endogenous

---

Fig. 20. *(opposite page) Xenopus* embryos express biochemically and immunochemically distinct human AChE. **(A)** Overexpression of rHAChE in developing embryos: HS/detergent extracts of ACHE-E6-injected and uninjected embryos were prepared and assayed for AChE activity in the presence and absence of the selective inhibitor echothiophate (3.3×10^{-7}*M*). Endogenous AChE activity in injected embryos was calculated according to an algorithm assuming 90% inhibition of rHAChE and 20% inhibition of frog AChE at this concentration of inhibitor (inset). Bar graph displays the total AChE activity measured per microinjected embryo on various days postfertilization and the calculated activities attributable to rHAChE (dark shading) and endogenous frog AChE (light shading). White circles represent the total AChE activity measured in uninjected control embryos at the same time-points. Data represent averages of 4–6 embryos from the indicated number (N) of independent microinjection experiments ± SEM. Inset: Selective inhibition of rHAChE by echothiophate. Homogenates representing endogenous frog (Fr) or recombinant human (rH) AChE were assayed for activity following 40 min preincubation with the indicated concentrations of echothiophate. **(B)** Immunochemical discrimination between rHAChE and embryonic *Xenopus* AChE: Affinity purified AChE from ACHE-E6-injected *Xenopus* embryos (rH), control uninjected embryos (Fr), human brain (hBr), and erythrocytes (hEr) was subjected to denaturing gel electrophoresis and protein blot analysis. Each lane represents approx 20 ng protein, except rH, which contained only 6 ng. Note the complete absence of immunoreactivity with embryonic *Xenopus* AChE. Prestained mol-wt markers were from Bio-Rad (Hercules, CA) (right). Reprinted from Seidman et al. (1994) with permission from Raven Press.

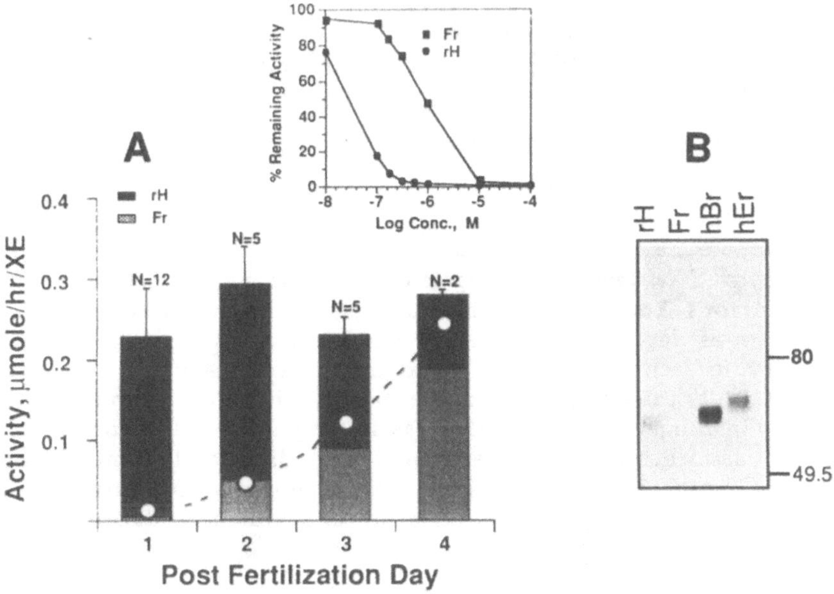

Fig. 20.

Xenopus AChE and heterologous AChE (Fig. 20A, inset), receding levels of rHAChE were observed for at least 4 d postfertilization. For the first 3 d, rHAChE accounted for >50% of the total measured AChE activity in microinjected embryos and resulted in a state of general overexpression compared to uninjected controls. By d 6 postfertilization, no heterologous enzyme could be detected in homogenates.

Sequential extractions with low salt (LS), low salt detergent (LSD), and high salt (HS) buffers revealed that approx 35% of rHAChE in ACHE-E6-injected embryos was associated with membranes, requiring detergent for solubilization (Table 7). Up to 33% of the catalytically active endogenous *Xenopus* enzyme in 3-d-old uninjected tadpoles appeared in the HS extractable fraction, presumably associated with the ECM. However, salt-soluble rHAChE remained primarily in the LS fraction.

Table 7
Hydrodynamic Properties of rHAChE
in ACHE-E6-Injected *Xenopus* Embryos[a]

Fraction	rH, d 1	rH, d 2	rH, d 3	Fr, d 3
LSS	57 ± 2	60 ± 4	53 ± 3	36 ± 5
DS	37 ± 2	34 ± 4	36 ± 3	31 ± 4
HSS	6 ± 2	5 ± 1	10 ± 1	33 ± 7

[a]Fertilized *Xenopus* eggs were microinjected with 1 ng ACHE-E6 DNA, cultured for 1–3 d and subjected to homogenization and subcellular fractionation as described in Experimental Methodologies in Chapter 2. rHAChE in each fraction (rH) was detected by EAIA using a specific MAb (101-1) raised against bovine brain AChE. Endogenous AChE activity in uninjected tadpoles (Fr) was determined by the standard colorimetric assay described in Experimental Methodologies. Percent enzyme activity in each fraction (average ± SEM) is shown for 3–5 groups of 3 embryos from a single microinjection experiment. Adapted from Seidman et al. (1994).

Apparently Normal Development of Acetylcholinesterase-Overexpressing Xenopus *Embryos*

The gross morphology and development of ACHE-E6-injected embryos appeared normal, except that injected embryos appeared somewhat smaller than uninjected controls (Fig. 19). Furthermore, the level of endogenous *Xenopus* AChE in ACHE-E6-injected tadpoles appeared lower than that observed in uninjected embryos (Fig. 19A). However, buffer-injected embryos displayed a similar slight retardation of growth and accumulation of endogenous AChE, indicating that these small perturbations probably reflect injection-associated injury. Nevertheless, gross motor function of microinjected embryos as evaluated by twitching and hatching on d 2, reflexive swimming on d 3, and free swimming on later days, appeared unimpaired compared to normal, uninjected controls. Microinjected tadpoles survived for up to 4 wk, showing no overt developmental handicaps.

Recombinant Human Acetylcholinesterase Is Immunochemically Distinct from Xenopus *Acetylcholinesterase*

In immunoblot analysis following denaturing gel electrophoresis, rHAChE comigrated with native human brain AChE, yielding a doublet band of approx 68 kDa (Fig. 20B). rHAChE was selectively recognized by a pool of MAbs raised against denatured bovine brain AChE (MAbs 132-1,2,3); no crossimmunoreactivity with embryonic *Xenopus* AChE was observed (Fig. 20B). rHAChE was similarly recognized in a species-specific manner by a variety of MAbs raised against native mammalian AChEs (Fig. 21).

Oligomeric Assembly of Recombinant Human Acetylcholinesterase in Xenopus *Embryos*

To examine the possibility that heterologous human AChE could undergo homomeric oligomeric assembly or interact with either catalytic or noncatalytic subunits of *Xenopus* AChE, sucrose density centrifugation was performed. At all time-points examined, rHAChE was observed exclusively as monomers sedimenting at approx 3.2 S, despite the concomitant accumulation of various multimeric forms of the endogenous *Xenopus* enzyme (Fig. 21). Endogenous *Xenopus* AChE appeared primarily as a dimer on d 2 postfertilization with globular tetrameric and asymmetric tailed forms appearing and increasing from d 3 onward (Fig. 21, insets; *see also* Lappin and Rubin, 1985). Superimposition of the gradients from control and ACHE-E6-injected embryos demonstrated that the normal developmental progression of *Xenopus* AChE oligomeric assembly was conserved in ACHE-E6-injected embryos despite the high excess of rHAChE monomers (Fig. 21).

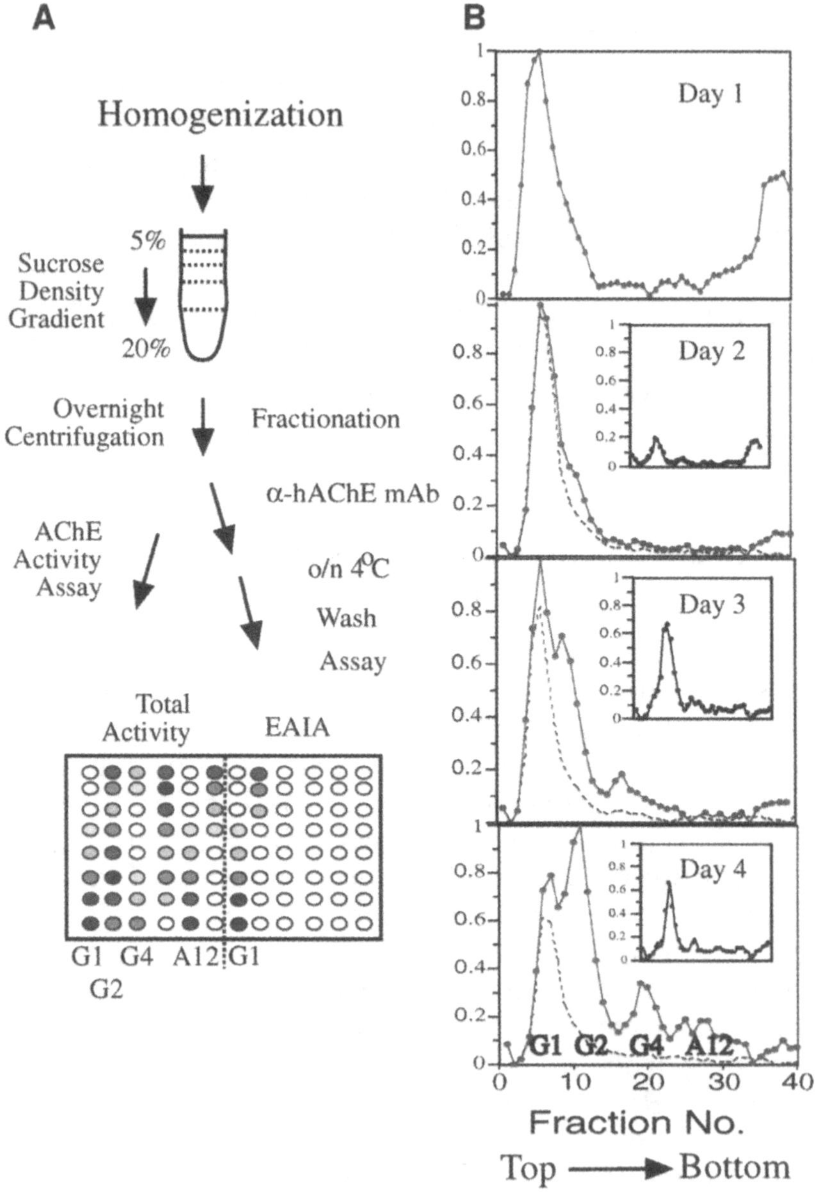
A
B
Homogenization
5%
Sucrose
Density
Gradient
20%
Overnight
Centrifugation
Fractionation
α-hAChE mAb
AChE
Activity
Assay
o/n 4°C
Wash
Assay
Total
Activity
EAIA
G1 G4 A12 G1
G2
Day 1
Day 2
Day 3
Day 4
G1 G2 G4 A12
Fraction No.
Top → Bottom

Fig. 21.

Characterization of a Human Acetylcholinesterase Gene Promoter in Xenopus Embryos

Human Acetylcholinesterase Gene Promoter Composition

Characterization of a 2.2-kb upstream sequence in the human ACHE revealed oligonucleotide motifs within the putative promoter region characteristic of binding sites for several known transcription factors including MyoD and SP1 (Ben Aziz et al., 1993a) (Fig. 14). Additional transcription-regulating motifs found in this sequence include the E-box enhancer octamer active in myogenesis, the cyclic adenosine monophosphate (cAMP) response element ATF/CREB, the EGR1 element, the AP-2 element, the Zeste sequence, the GAGA motif, the NF-kappa-B element, and both TATA and CCAAT sites. The ability of this upstream sequence to drive transcription was assessed by linking it to the AChE-E6 coding sequence to generate a human AChE promoter-reporter construct (HpACHE) and injecting this vector into oocytes and fertilized eggs.

Fig. 21. *(previous page)* rHAChE remains unassembled in *Xenopus* embryos. HS/detergent extracts representing 2 embryos were subjected to sucrose density centrifugation and EAIA. **(A)** Experimental strategy as described in Experimental Methodologies in Chapter 2. **(B)** Figure represents total AChE (solid line) and immunoreactive rHAChE (dotted line) from ACHE-E6-injected embryos 1–4 d postfertilization. rHAChE appeared exclusively as a peak representing monomeric AChE (approx 3.2 S) at all time-points. The distribution of *Xenopus* AChE was unchanged compared to controls (insets). Arrow marks position of bovine liver catalase (11.4 S). Insets: AChE molecular forms in control uninjected embryos scaled to the total activity levels observed in DNA-injected embryos (*see* Fig. 11). Note that monomeric AChE is essentially undetectable in control embryos. Reprinted from Seidman et al. (1994) with permission from Raven Press.

Transcription from the Human Acetylcholinesterase Gene Promoter in Xenopus *Detected by RT-PCR*

HpACHE was injected into oocytes or embryos in either its full form (S) or as one of several truncated derivatives (K, X, or P) (Fig. 22). Transcriptional activity was evaluated by RT-PCR using human AChEcDNA-specific primers. ACHE-E6 behind the CMV promoter served as control. Only RNA from oocytes and embryos injected with the complete HpACHE gave rise to the expected 275-bp PCR product (Fig. 22). Thus, the intact promoter sequence was sufficient to support transcription in *Xenopus*, but not the TATA and CAAT boxes or any of the other transcription factor binding sites alone. Naturally, the mere existence of a transcription binding site within a promoter does not necessarily predict its function in a microinjection assay.

Microinjected Embryos Utilize Correct 5' Splice Site

To locate functional splice sites within the upstream genomic sequence, several consensus splice site motifs were tested by RT-PCR of RNA extracted from various human tissues. Chimeric primer pairs were designed in which the 12 5' nucleotides of each upstream primer (+) terminated at one of the putative 5' splice sites and the 3'-terminal 5 or 6 nucleotides corresponded to the common putative acceptor site (Fig. 23). The downstream primer (–) was located within the coding sequence. With this strategy, a single 5' splice site was identified which delimited a 1.5-kb intron (I1) in the human ACHE gene. When RNA from HpACHE-injected oocytes and embryos was subjected to RT-PCR using the chimeric primer pair representing the native splicing option, the identical 400-bp fragment was produced, indicating that the splice-recognition signal present in this human gene is similarly recognized in *Xenopus*.

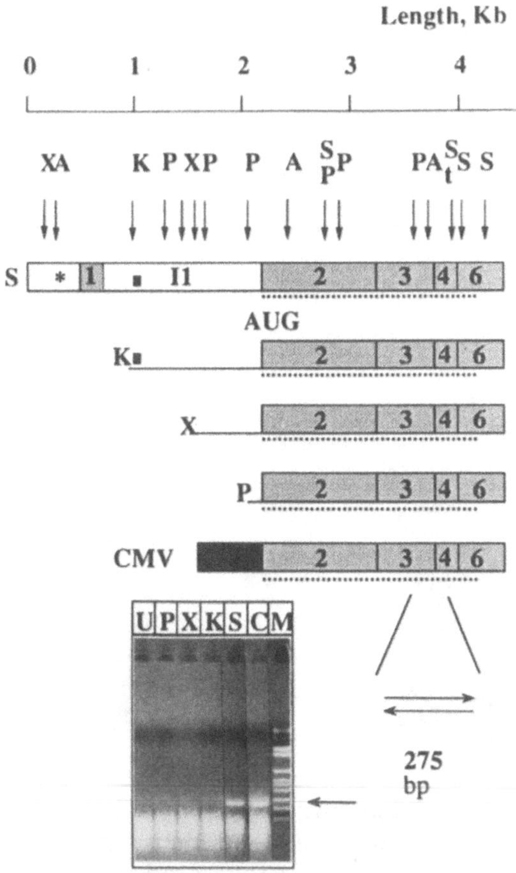

Fig. 22. Expression of the human ACHE promoter in *Xenopus*. Injected DNAs included a 2.2-kb upstream sequence in its full form (S) or following digestion by Kpn (K), Xho (X), or Pvu (P) linked to the common AChE coding exons E2, E3, E4, E6. ACHE-E6 carrying the coding sequence under control of the CMV promoter served as control. Putative TATA box and cluster of potential transcription factor binding sites are marked by a solid box and asterisk, respectively. Open reading frames are noted by dotted underlines. Inset: RT-PCR analysis of RNA extracted from *Xenopus* oocytes 2 d following injection with the noted linearized constructs. Residual injected DNA was eliminated with DNase I. Lanes: U, uninjected oocytes; P, X, K, S, injected construct; C, CMV; M, DNA size marker VI (Boehringer Mannheim). PCR products indicated by the arrow. Reprinted from Ben Aziz-Aloya et al. (1993a) with permission from National Academy of Sciences.

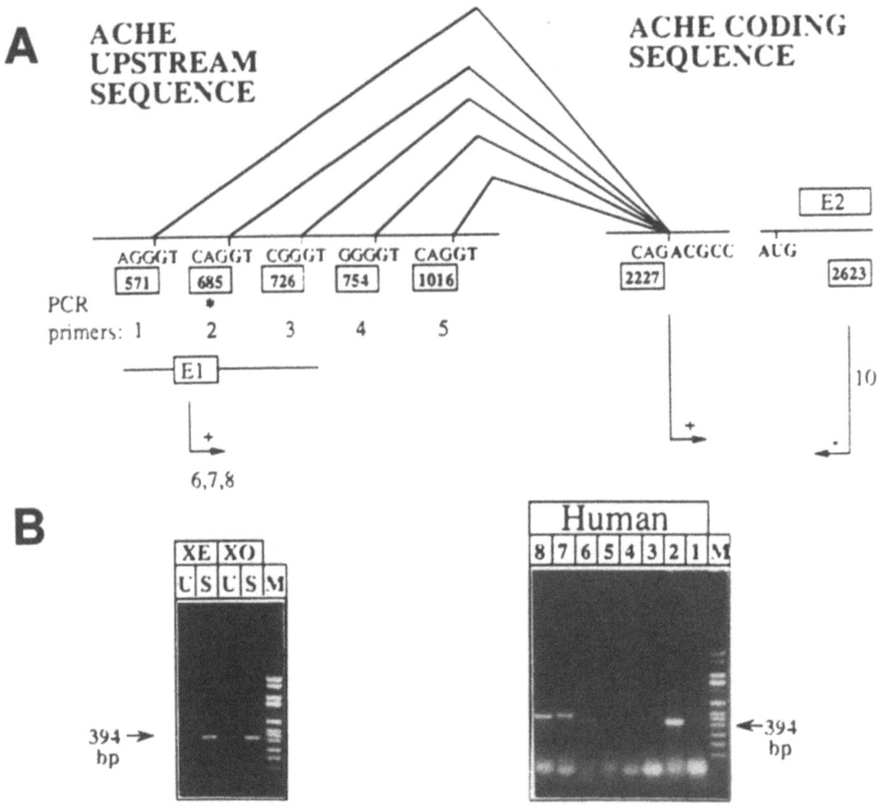

Fig. 23. Correct 5' splicing of human ACHE mRNA in *Xenopus*. **(Upper)** Potential 5' splice sites (consensus G/GT) were identified in the ACHE upstream region at the noted positions. A potential 3' acceptor site was identified at position 2226, 21 nucleotides upstream from the AUG start site. Chimeric downstream-oriented primers (+) included 15–21 nucleotides from the potential 5' donor site and 5–6 nucleotides from the putative 3' receptor site. Primers were computer-designed to recognize only their entire intact counterparts, which would only be present if splicing occurred. All were tested with a single upstream oriented primer from coding exon E2 (primer no. 10; Ben Aziz-Aloya et al., 1993a). The asterisk indicates the experimentally confirmed active splice site. **(Lower right inset)** Activity of primer no. 2 and two closely positioned upstream primers (nos. 7 and 8; Ben Aziz-Aloya et al., 1993a) in promoting RT-PCR using RNA derived from adult human brain. **(Lower left inset)** Activity of primer no. 2 in amplification of RNA extracted from both *Xenopus* embryos (XE) and oocytes (XO) injected with HpACHE, indicating recognition of the native human splice boundaries in *Xenopus*. Reprinted from Ben Aziz-Aloya et al. (1993a) with permission from National Academy of Sciences.

Unique Properties of an Alternative Acetylcholinesterase Expressed in Xenopus *Embryos*

To study the biological implications of 3' alternative splicing in the ACHE, an additional CMV driven ACHE plasmid potentially encoding both the GPI-linked erythrocyte and putative readthrough AChE forms was constructed (ACHE-I4/E5) (*see* Figs. 13 and 15). When ACHE-I4/E5 was introduced into fertilized *Xenopus* eggs, the levels and timing of rHAChE expression were similar to those observed with ACHE-E6 (Fig. 24A). As with ACHE-E6-injected embryos, overall development of ACHE-I4/E5-injected embryos appeared normal through gastrulation, neurulation, hatching, and the acquisition of motor function. Moreover, RT-PCR performed with primers specific for XmyoD, expressed in muscle (Hopwood et al., 1991) and *Xenopus* GATA-2 expressed in developing erythrocytes (Zon et al., 1991) indicated that no global changes in the levels of RNA encoding these proteins were associated with overexpression of either ACHE-I4/E5 or ACHE-E6 DNAs. In sucrose density gradient centrifugation, a single peak at 3–5S was observed for ACHE-I4/E5-derived AChE, indicating a monomeric configuration for this recombinant enzyme, consistent with the lack of a C-terminal cysteine as predicted by the cDNA sequence.

Despite the similarities between rHAChEs derived from ACHE-I4/E5 and ACHE-E6 in the embryos, several striking differences were observed. AChE activity from embryos injected with ACHE-E6 consistently partitioned into both the low salt (55%) and low salt detergent fractions (35%) (Fig. 24B). In contrast, activity from embryos injected with ACHE-I4/E5 was 85–90% solubilized in the low salt step. In nondenaturing PAGE, catalytically active AChE from ACHE-I4/E5-injected embryos migrated as a triplet band and ran significantly faster than the single band representing AChE from several recombinant and natural

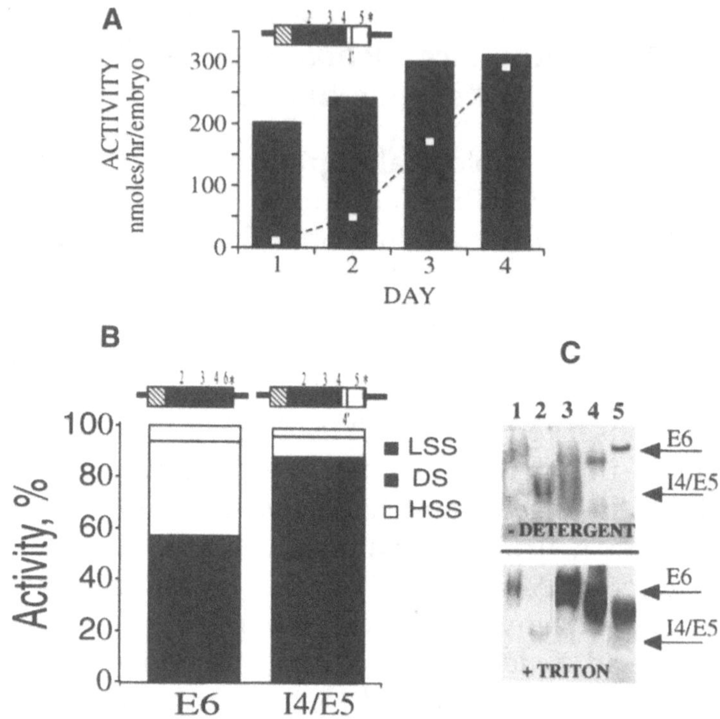

Fig. 24. Expression of ACHE-I4/E5 DNA in microinjected *Xenopus* embryos. **(A)** Transient expression: Embryos injected with approx 1 ng ACHE-I4/E5 DNA under control of the CMV promoter were raised for 1–4 d at 20°C. Columns represent total AChE activity measured in homogenates from DNA-injected embryos. Squares represent endogenous AChE activity measured in age-matched control uninjected embryos. Note the developmentally regulated increase in endogenous AChE levels and the concomitant decline in the relative extent of overexpression from d 1–4. Data represent average of three separate groups of 4 embryos from a single microinjection experiment. **(B)** Differential solubility: Sequential extractions of 10 one-d-old DNA-injected embryos into LSS, LSD, and HSS fractions was performed. Columns represent the relative fraction of the total summed activities extracted at each step for a representative experiment. Endogenous *Xenopus* AChE activities were considered negligible and ignored. Note the predominantly salt-soluble nature of the ACHE-I4/E5 product(s) as compared with rHAChE as encoded by ACHE-E6. **(C)** Distinct electrophoretic mobility: Nondenaturing gel electrophoresis was performed in the presence or absence of 0.5%

sources (Fig. 24C). When electrophoresis was performed in the absence of detergent, no significant shift in the migration of ACHE-I4/E5-derived bands was observed. Together, these observations implied that, despite the inclusion of exon E5 in ACHE-I4/E5, rHAChE derived from this plasmid in *Xenopus* represented a non-GPI-linked, nonamphiphilic species.

A Novel AChE mRNA Species Characterized in Xenopus

ACHE-I4/E5 potentially encodes both an mRNA encoding the GPI-linked erythrocyte AChE by splicing out of I4, and/or an unspliced mRNA coding for the putative readthrough AChE. However, the soluble character of the protein expressed in ACHE-I4/E5-injected embryos strongly implied that no GPI moiety was present on this heterologous AChE. The observed properties could be explained by either of two possibilities: the dominant heterologous mRNA present in the embryos is unspliced readthrough mRNA, or incomplete processing or instability of the E5 peptide such that the hydrophobic C-terminal peptide is removed, but the GPI anchor is not appended. To evaluate the nature of the AChE mRNAs generated in microinjected *Xenopus* embryos, total RNA extracted from

---

Triton X-100. Lanes 1 and 2—ACHE-E6 and ACHE-I4/E5—derived products, respectively, as expressed in *Xenopus* embryos (1 d postfertilization); lane 3, purified human erythrocyte AChE; lanes 4 and 5, rHAChE-E6 as expressed in human 293 cells and *E. coli*, respectively. Note that ACHE-I4/E5 product(s) migrate as three bands and significantly faster than all the other AChE forms both in the presence and absence of detergent. No band representing endogenous *Xenopus* AChE could be detected at this time-point. The smearing of activity in lane 3 of the no detergent (–) gels probably represents aggregation of the hydrophobic GPI-linked AChE erythrocyte dimer in the absence of Triton. Reprinted from Seidman et al. (1995) with permission from American Society for Microbiology.

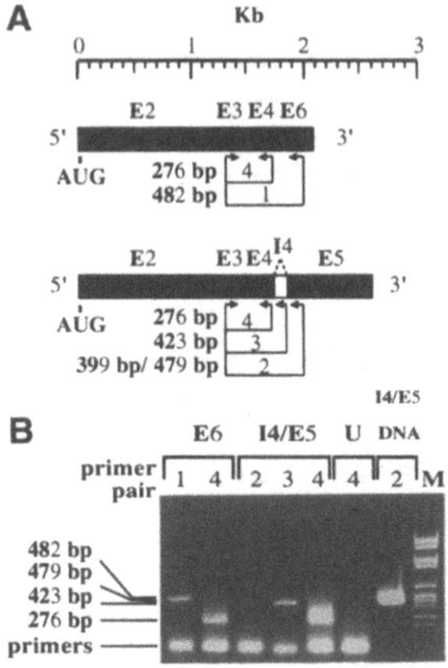

Fig. 25. ACHE-I4/E5 DNA gives rise to a truncated readthrough mRNA. **(A)** Schematic representation of possible ACHE mRNA products: Schemes present ACHE-E6 **(upper)** and ACHE-I4/E5 **(lower)** mRNAs; I4 may be spliced out of the primary ACHE-I4/E5 transcript (displayed by dashed triangle), or retained in the mature mRNA. The expected product lengths of PCR primer pairs 1–4 are designated to the left of the arrows. **(B)** RT-PCR: Reverse transcription and PCR were performed on total RNA from 1-d-old *Xenopus* embryos injected with the ACHE-E6 (E6) or ACHE-I4/E5 (I4/E5) DNA constructs. Uninjected embryos (U) served as control. Control reactions, without RT, did not yield amplification products, proving the absence of contaminating DNA sequences (not shown). DNA-PCR analysis performed with the ACHE-I4/E5 DNA construct (I4/E5 DNA) confirmed the ability of primer pair number 2 to yield a PCR product. M, markers. Reprinted from Seidman et al. (1995) with permission from American Society for Microbiology.

ACHE-E6-injected, ACHE-I4/E5-injected or control, uninjected embryos was subjected to RT-PCR with I4-, E4-, E5-, and E6-specific primers (Fig. 25).

When RNA from ACHE-I4/E5-injected embryos was subjected to RT-PCR, both the invariant exon E4 and the intronic sequence I4 were detected. However, the E5-specific primer pair repeatedly failed to generate either the 479-bp fragment representing the full-length readthrough mRNA or the 399-bp fragment representing spliced E5-bearing mRNA (Fig. 25). A parallel reaction using control plasmid DNA and the identical E5-specific primers did yield the 479-bp band, however, validating the efficacy of this primer pair (Fig. 25). RT-PCR performed on RNA extracted from ACHE-E6-injected embryos indicated the presence of both E4 and E6 (Fig. 25). These data therefore indicated that rHAChE activity induced by ACHE-E6 reflected the complete brain and muscle AChE, whereas heterologous AChE activity produced in *Xenopus* from ACHE-I4/E5 was derived from a truncated readthrough ACHE mRNA.

Tissue-Specific Management of Human Acetylcholinesterases Derived from Alternative AChE mRNAs

Whole-Mount Cytochemical Staining Reveals Tissue-Specific Accumulations of Acetylcholinesterase

To examine the ability of alternative splicing to account for tissue-specific accretion of AChE, in vitro fertilized *Xenopus* eggs were microinjected with 1 ng ACHE-E6 or ACHE-I4/E5 DNA. The resultant embryos were raised for 2–3 d, fixed, and stained in whole mount for catalytically active enzyme using the cytochemical activity staining method of Karnovsky and Roots (1964) (*see* Appendix VI for a detailed protocol). Following injection of ACHE-E6 encoding the brain and muscle form of AChE, 2-d-old tailbud embryos displayed conspicuous overexpression of AChE in the developing myotomes (Fig. 26). Myotomal overexpression was primarily observed as pronounced longitudinal staining along the plane of the muscle fibers between the vertical bands representing natural accumulations of AChE at the intersomitic junctions. However,

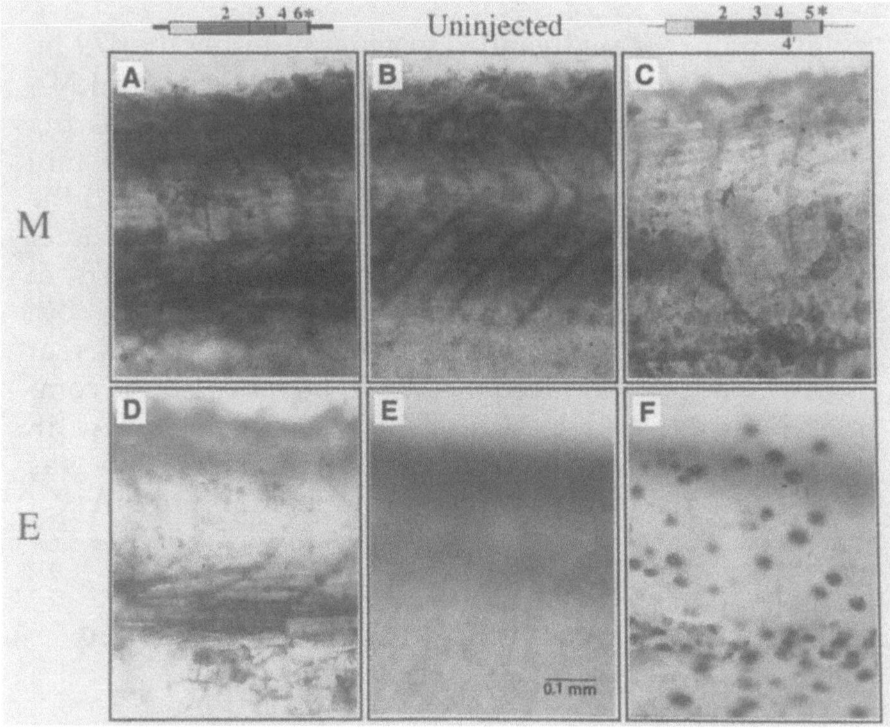

Fig. 26. Alternative AChE mRNAs dictate tissue-specific accumulation of AChE. ACHE-E6-injected, ACHE-I4/E5-injected or control uninjected *Xenopus* embryos were raised for 2 d at 19°C, fixed, stained for catalytically active AChE, and examined. Schematic representations of the microinjected DNAs, which gave rise to the staining pattern, exhibited in each micrograph are shown. M, micrographs taken at the focal depth of muscle; E, micrographs taken at the focal depth of epidermis. Rostral (anterior) truck myotomes from an ACHE-E6-injected embryo **(A)**. Note, in addition to the natural intersomitic vertical bands of brown stain, the dark—almost black—horizontal bands of staining parallel to the plane of the myotomal muscle fibers. No detectable staining was noted in the epidermal layer, although the heavily stained myotomes beneath the skin could be seen when a higher focal plane was examined **(D)**. Mid-body view of an ACHE-I4/E5 expressing embryo displaying punctuated epidermal staining **(F)**. No enhancement of AChE activity in the myotomes of these embryos was detected **(C)**. Center panels, uninjected control embryo. Note the accumulation of reaction product at the junctions flanking the individual somites and the very light staining along the horizontal plane of the muscle fibers **(B)**. No staining is observed in epidermis **(E)**. Reprinted from Seidman et al. (1995) with permission from American Society for Microbiology.

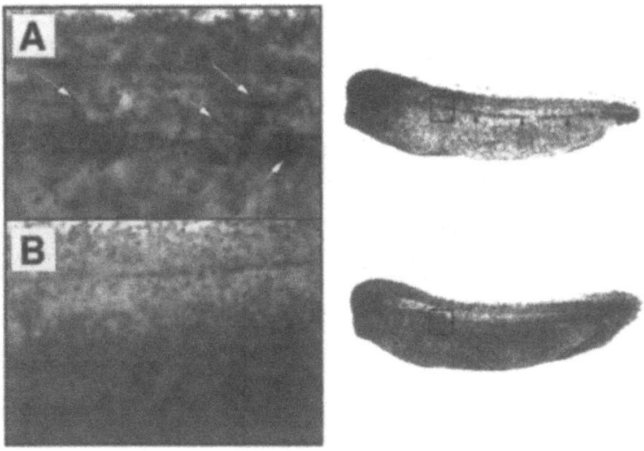

Fig. 27. Whole-mount immunocytochemical staining reveals accumulation of human AChE *Xenopus* muscle. Two-day-old *Xenopus* embryos were raised and stained for immunoreactive human AChE using a pool of MAbs specific for human enzyme as described in Experimental Methodologies in Chapter 2. **(A)** Low magnification image of an ACHE-E6 injected embryo overexpressing recombinant human AChE along a longitudinal band within the somites. Arrows mark sites of immunoreactive material. **(B)** A close-up view of a rostral trunk somite from an AChE-overexpressing (A) and control uninjected (B) embryo. Note the similarity between the immunocytochemical and AChE activity staining patterns in muscle (Fig. 26).

myotomal overexpression of AChE was clearly mosaic, varying in intensity within and between individual somites. Uninjected control embryos displayed the characteristic staining at the intersomitic junctions, but only faint staining within individual myotomes (Fig. 26). Whole-mount immunocytochemical staining with a pool of MAbs raised against mammalian AChE verified the accumulation of recombinant human enzyme in somitic muscle cells (Fig. 27).

In contrast to the striking accumulation of ACHE-E6-derived AChE in myotomes, we did not observe any notable enhancement of staining in myotomes of embryos injected with ACHE-I4/E5. Rather, we noted pronounced punctuated staining of the epidermis, which was never seen in uninjected embryos (Fig. 25C). Intersomitic staining was unaffected by expression of ACHE-I4/E5. Although limited epidermal staining was occasionally observed in ACHE-E6-injected embryos, this phenomenon appeared restricted to sites of particularly high myotomal expression and was considerably less well defined. These observations indicated that AChE derived from ACHE-E6 DNA was specifically accumulated in muscle, whereas AChE derived from ACHE-I4/E5 was uniquely targeted to the epidermis.

Electron Microscope Analysis Reveals Subcellular Compartmentalization of Human Acetylcholinesterase in Xenopus *Muscle*

To examine the myotomal overexpression of ACHE-E6 in detail, an ultrastructural analysis was undertaken at the electron microscope level. Longitudinal and transverse sections from rostral trunk somites (Fig. 28) revealed clearly discernible myofibers 2 d postfertilization in both ACHE-E6-injected and control, uninjected embryos (Fig. 29). By d 3, both groups displayed significant increases in their numbers of myofibrillar elements and in maturation of the sarcoplasmic reticulum (SR) (Fig. 29). The subcellular localization of nascent AChE in transgenic and control embryos was determined using the cytochemical activity staining method of Karnovsky (1964). In both the experimental and control groups, crystalline deposits of electron dense reaction product were observed primarily in association with myofibrils, amidst the myofilaments and within the SR (Figs. 29 and 30). Various organelles, including the nuclear membrane, free and bound polyribosomes, golgi, and sometimes mitochondria, were also observed to be stained.

At d 2 postfertilization, staining in DNA-injected embryos was conspicuously more pronounced than that observed in uninjected controls, both in the quantity and intensity of reaction product (Fig. 29). However, variability

Stage	I	II	III	IV	V	VI
Size, uM	50-300	300-450	450-600	600-1000	1000-1200	1200
Morphology	Transluscent Oval	Opaque White, Oval	Even light brown, Round	Uneven light brown	Clearly defined Hemispheres	Equatorial belt
rRNA synthesis		+	+	+	+	+
mRNA synthesis	+	+	+	+	+	+
Mitochondrial production and organization	+	+	+	+	+	+
Vitellogenesis			+	+	+	

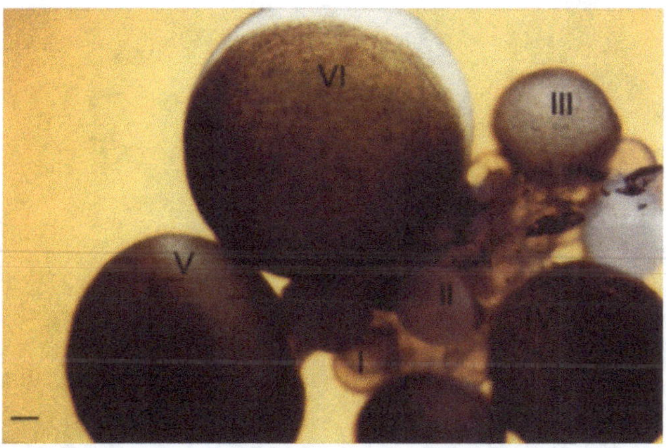

Plate 1 (Fig. 1; *see* full caption on p. 3 and discussion in Chapter 1). Stages of *Xenopus* oogenesis. Stage 5 or 6 oocytes are those commonly used for microinjection experiments.

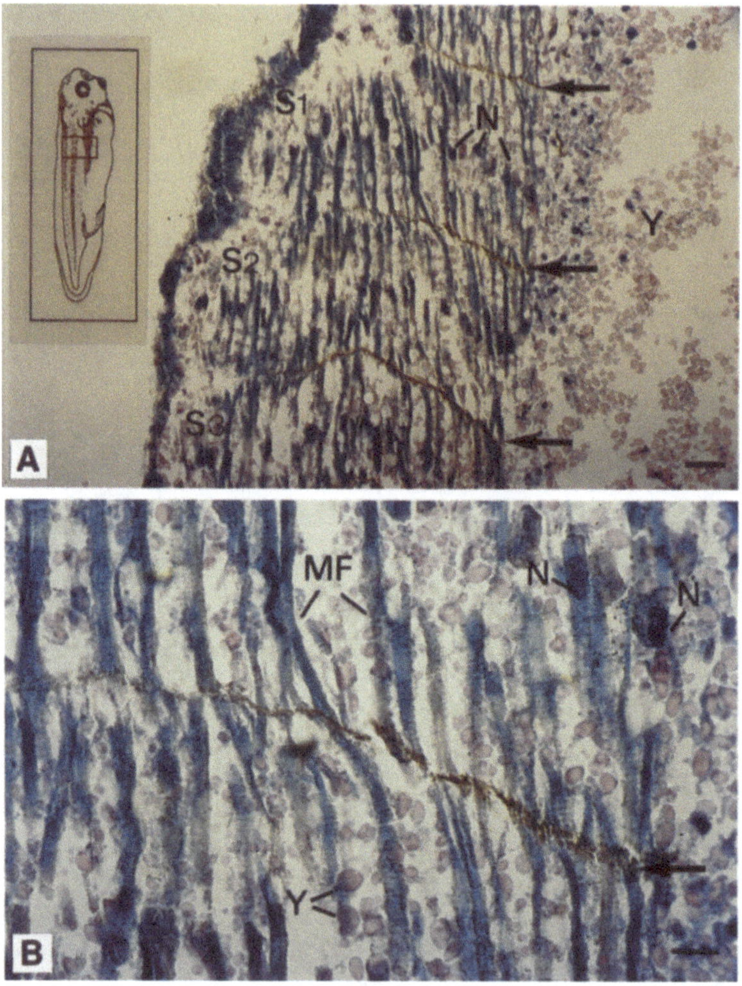

Plate 2 (Fig. 7; *see* full caption on p. 17 and discussion in Chapter 1). Intersomitic junctions of a stage 35/36 *Xenopus* embryo revealed by cytochemical staining for catalytically active AChE. Counterstain was hematoxylin-eosin-phosphomolybdic acid/light green stain. MF, myofibers; N, nuclei; S, somite; Y, yolk platelets (magnification **[A]** ×512 and **[B]** ×1600).

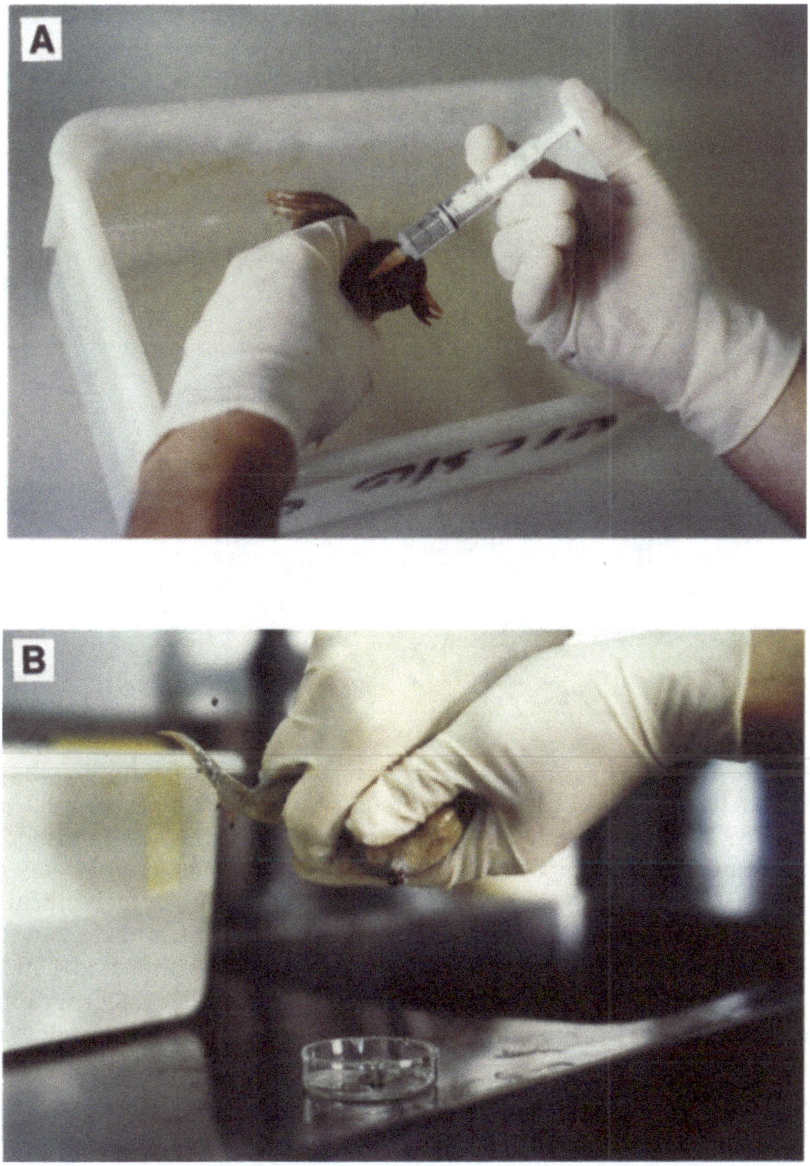

Plate 3 (Fig. 16; *see* full caption on p. 80 and discussion in Chapter 2). Preparing *Xenopus* eggs for in vitro fertilization. Female *Xenopus* frogs are primed with HCG **(A)** and mature fertilizable eggs are collected 16–18 h later **(B)**.

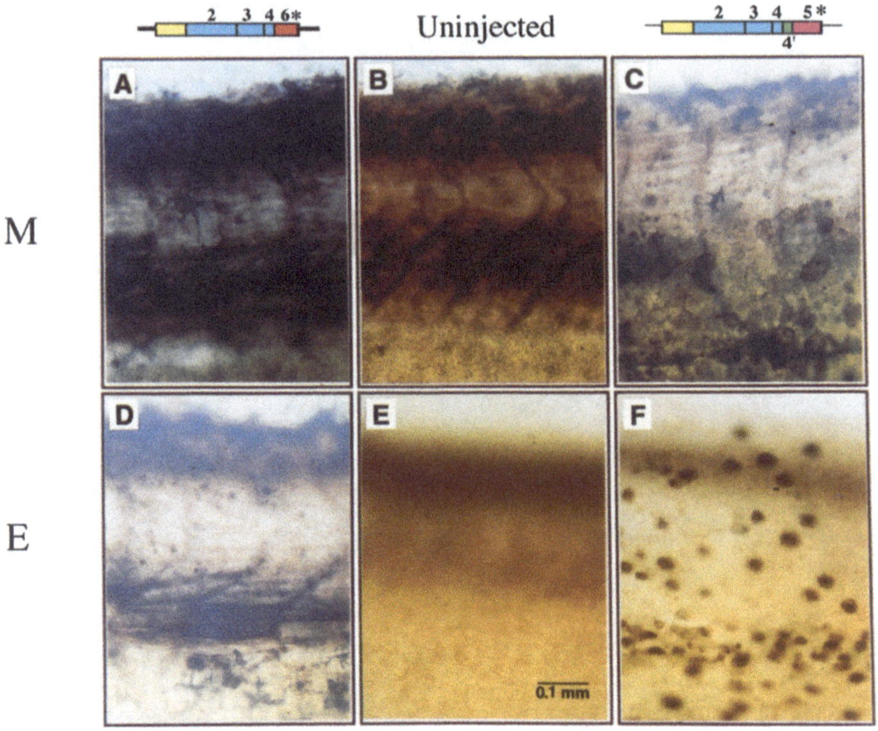

Plate 4 (Fig. 26; *see* full caption on p. 108 and discussion in Chapter 3). Alternative AChE mRNAs dictate tissue-specific accumulation of AChE in *Xenopus* embryos. In vitro-fertilized *Xenopus* eggs microinjected with DNA encoding alternative AChE forms downstream of the CMV promoter were raised for 2 d at 19°C, fixed, cytochemically stained for catalytically active AChE in whole mount, and cleared for microscopy. Schematic representations of the microinjected DNAs appear above. Uninjected embryos served as control. M, micrographs taken at the focal depth of myotomal muscle; E, micrographs taken at the focal depth of epidermis. Reprinted from Seidman et al. (1995) with permission from American Society for Microbiology.

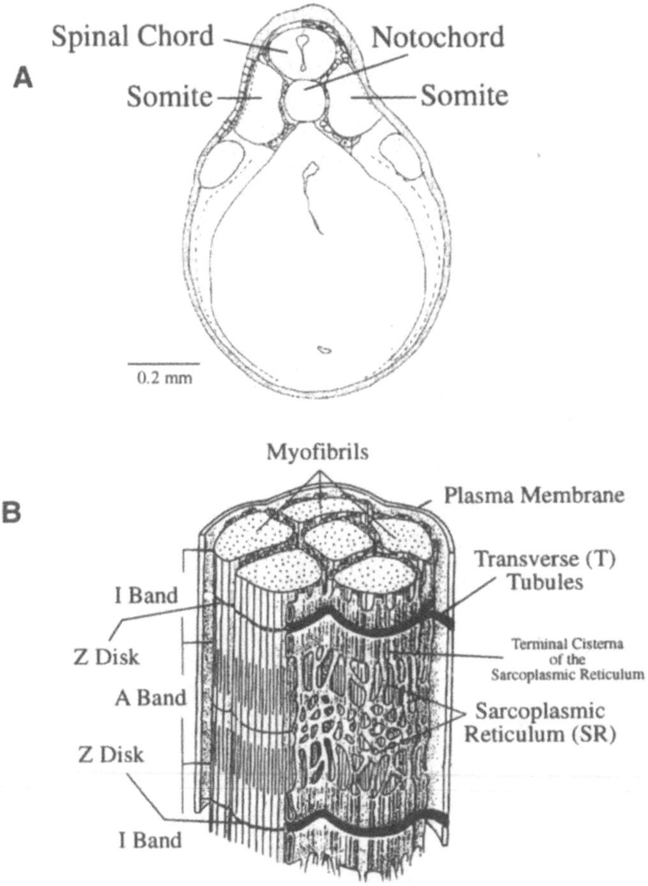

Fig. 28. Schematic cross-sectional representation of 2-d-old *Xenopus* embryo. **(A)** A d-2 tailbud embryo is shown in cross-section at the level at which the following electron micrograph studies were performed. Adapted from Hausen and Riebesell (1991) with permission from Springer Verlag. **(B)** Schematic representation of a vertebrate skeletal muscle fiber, illustrating the organization of the muscle fibrils with their associated sarcoplasmic reticulum and T-tubule systems within which both native *Xenopus* and rHAChE were localized in microinjection experiments.

was observed between tissue blocks, probably reflecting the mosaic expression of the injected DNA and/or variability in the efficiency of expression between embryos. In longi-

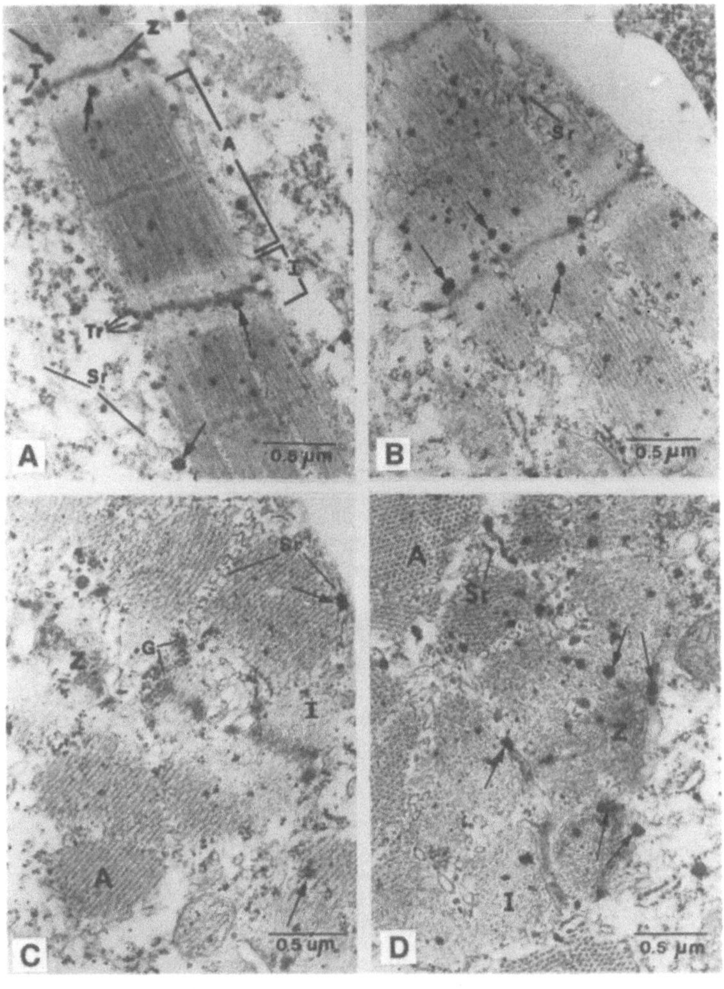

Fig. 29. Human AChE in myotomes of 2-d-old microinjected *Xenopus* embryos. *Xenopus* eggs were injected with ACHE-E6, incubated for 2 d at 17°C, fixed, stained for catalytically active AChE, and prepared for electron microscopy as described in Experimental Methodologies in Chapter 2. Uninjected embryos from the same fertilization served as controls. Arrows mark reaction product. **(A)** Longitudinal section of myotome from uninjected control embryo. **(B)** Myotome from ACHE-E6-injected embryo. **(C)** Control myotome in transverse section. **(D)** Transverse section from ACHE-E6-injected embryo. Note the increased number, size, and density of crystals in sections from DNA-injected embryos vs controls within the same subcellular compartments. A, A band; G, glycogen particles; I, I band; Sr, sarcoplasmic reticulum; T, T tubules; Tr, triad; Z, Z disk. Reprinted from Seidman et al. (1994) with permission from Raven Press.

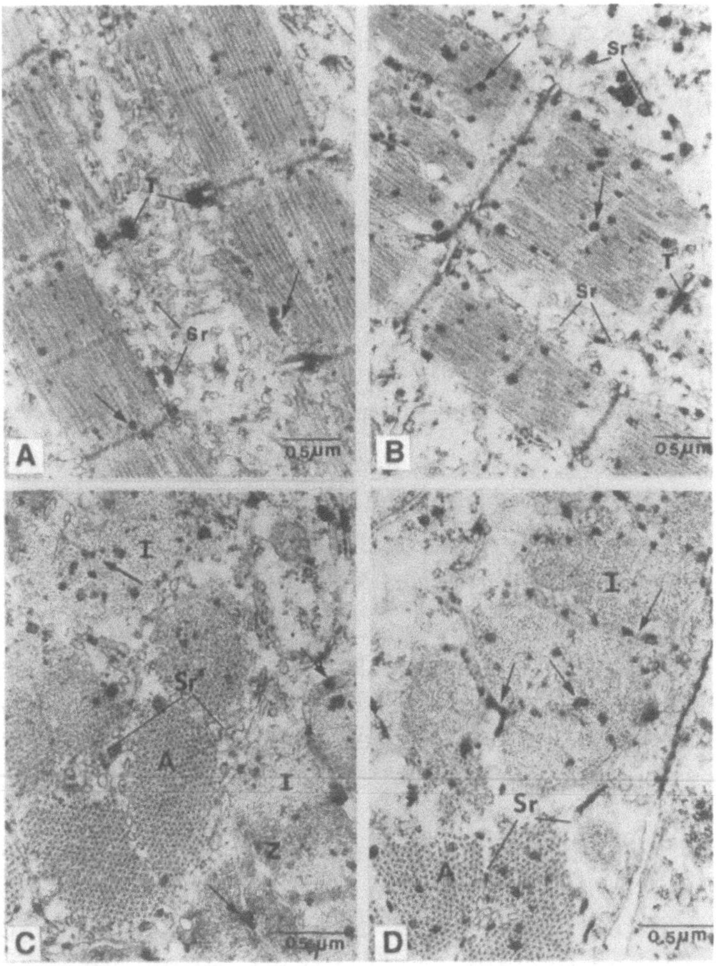

Fig. 30. Myotomal overexpression of AChE persists to d 3; as in Fig. 29, except that embryos were analyzed after 3 d incubation. Note the developmental increases in myotomal AChE in both control uninjected **(A,C)** and ACHE-E6-injected sections **(B,D)**, especially within the Sr and T tubules. Reprinted from Seidman et al. (1994) with permission from Raven Press.

tudinal sections from ACHE-E6-injected embryos, staining appeared to be concentrated at the I band of myofibers, particularly around the triad marking the intersection of the SR and T-tubule systems (schematically presented in Fig. 28). In contrast, the sparse staining observed in control sections appeared randomly distributed.

By d 3 postfertilization, the general staining intensity in both groups had significantly increased. However, the observable differences between the groups were less dramatic. Cross-sections revealed especially prominent staining within the SR (Fig. 30A,B). Strong staining was now observed at both the A and I bands and, for the first time, within the T-tubules (Fig. 30C,D). Overall, myotomes from 2-d-old ACHE-E6-injected embryos resembled d-3 uninjected, control myotomes in staining incidence and intensity.

Accumulation of Acetylcholinesterase in Neuromuscular Junctions of DNA-Injected Xenopus *Embryos*

Since overexpressed AChE was observed within its natural subcellular compartments in muscle, we asked whether heterologous AChE could be localized within NMJs as well. In NMJs from 2-d-old ACHE-E6-injected embryos, up to 10-fold overexpression of catalytically active AChE was observed compared with NMJs from control embryos (Ben Aziz-Aloya et al., 1993a). To examine the persistence of this state and its implications for synaptic ultrastructure, both cytochemically stained and closely appositioned unstained NMJs from 3-d-old injected and control embryos were studied (Fig. 31 and Table 8). In the ACHE-E6-injected group, 72% of the postsynaptic membrane length (stained length [SL] and postsynaptic length [PSL]) was stained, on average, for active AChE. In contrast, only 22% of the PSL was stained in controls. Moreover, the total area (SA) covered by reaction product was approx fourfold greater in NMJs from ACHE-E6-injected embryos than those from controls. In addition, the staining observed in NMJs from injected embryos was considerably more intense than that displayed by control NMJs, forming large black accumulations of reaction product as

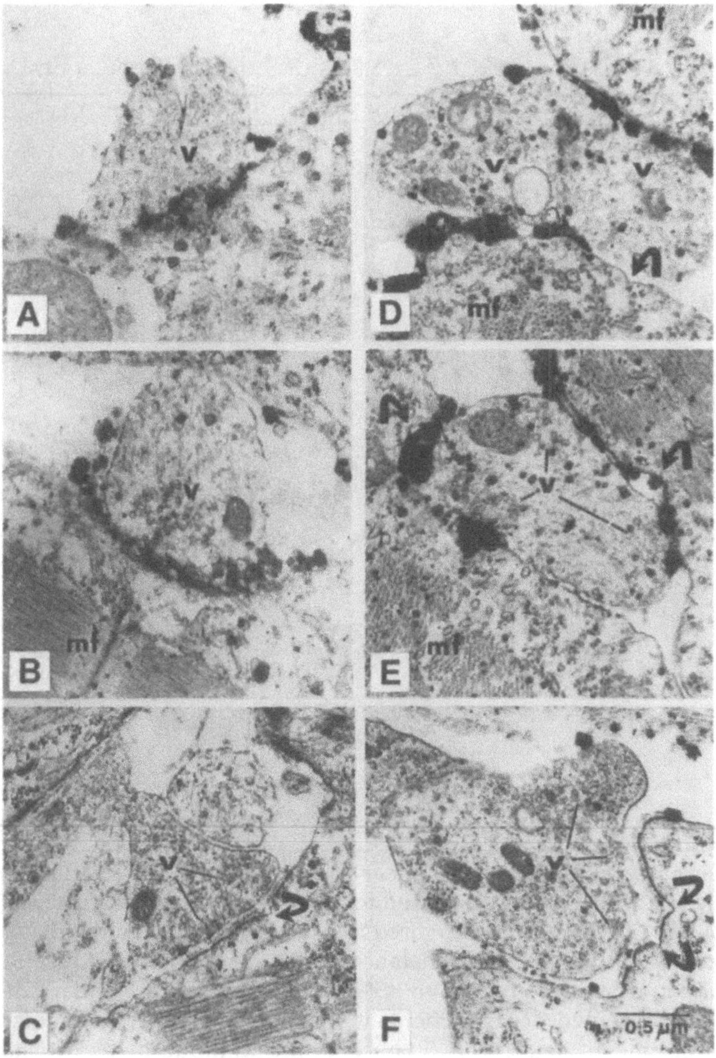

Fig. 31. Accumulation of AChE in NMJs of 3-d-old *Xenopus* embryos overexpressing human AChE. Electron micrographs were prepared as in Fig. 29. Cross-sections of two cytochemically stained synapses are presented from an uninjected control **(A,B)** and ACHE-E6-injected **(D,E)** embryo. Note the particularly high density staining in areas directly opposite nerve-terminal zones enriched in neurotransmitter vesicles (V). **(C,F)** Representative unstained NMJs from a control (C) and a ACHE-E6-injected (F) embryo. The synapse presented in B represents the highest degree of staining observed in a control section. Arrows, postsynaptic folds, mf, myofibril; v, presynaptic neurotransmitter vesicles. Reprinted from Seidman et al. (1994) with permission from Raven Press.

Table 8
Overexpression of AChE in NMJs of 3-d-Old *Xenopus* Embryos[a]

Morphometric parameter	ACHE-E6, $N = 8$	Uninjected, $N = 8$	ACHE-E6-uninjected	p
PSL, μm	2.64 ± 0.58	1.88 ± 0.93	1.4	<0.01
SL, μm	1.85 ± 0.33	0.58 ± 0.22	3.2	<0.002
SL/PSL, %	72.0 ± 15.0	22.0 ± 19.0	3.3	<0.002
SA, μm²	0.37 ± 0.09	0.084 ± 0.38	4.4	<0.002

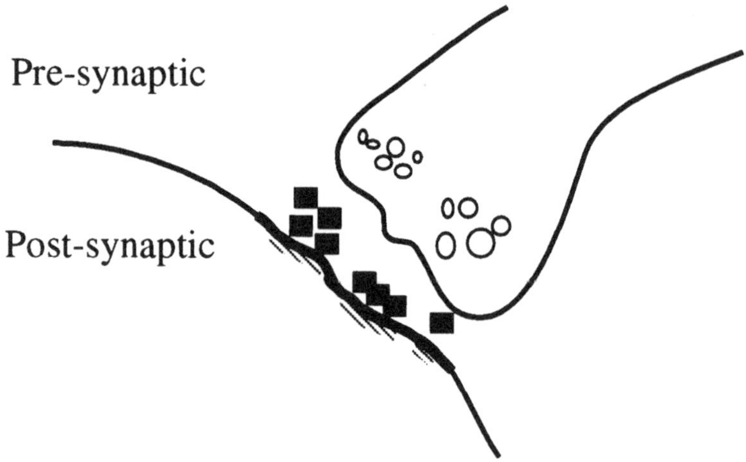

[a]Eight representative synapses from ACHE-E6-injected or control-uninjected embryos were assessed for postsynaptic membrane length in μm (PSL), the sum total length covered by reaction product in μm (SL), the fraction of nerve–muscle contact distance displaying reaction product (SL/PSL), and the total stained area in μm² (SA). Average values ± standard deviation (SD) are presented. Measurements were performed on electron micrograph photographs as detailed in Experimental Methodologies in Chapter 2. Adapted from Shapira et al. (1994).

opposed to the lighter, more diffuse staining observed in controls (Fig. 31 A,B,D,E).

Ultrastructural features of NMJs from injected and uninjected embryos were best discerned in unstained synapses. NMJs from control embryos generally appeared smooth and relatively undeveloped, with up to two secondary folds of the postsynaptic membrane, and a single nerve–muscle contact (Fig. 31C). In contrast, NMJs from ACHE-E6-injected embryos displayed an average of three secondary folds and 1–3 discrete contacts between pre- and postsynaptic membranes (Fig. 31F). Furthermore, the average postsynaptic membrane length in NMJs from ACHE-E6-injected embryos was 30% larger and considerably less variable than that measured in control embryos (SL) (Table 8). Moreover, the distance across the synaptic cleft was both larger and more variable in injected embryos than in controls (129 ± 72 nm vs 94 ± 23 nm; $N = 14$). These observations implied a correlation between synaptic overexpression of AChE and enhanced NMJ development. Therefore, to speed up embryonic development and allow NMJ development to coincide more fully with the window of highest rHAChE expression (d 1–2), the next experiments were performed at 19–21°C.

C-Terminal Peptide Dictates Synaptic Localization of Heterologous Acetylcholinesterase

The mean AChE-stained cross-sectional area of NMJs from ACHE-E6-injected embryos reached three or four times that observed in NMJs from uninjected embryos (0.33 ± 0.29 μm^2 vs 0.08 ± 0.09 μm^2) (Fig. 32). In contrast, NMJs from embryos injected with ACHE-I4/E5 displayed only a minor increase compared with controls (0.14 ± 0.03 μm^2) (Fig. 32). Transverse sections of myofibers displayed a parallel pattern of highly intensified staining in myotomes of ACHE-E6-injected embryos compared with both ACHE-I4/E5-injected and uninjected controls.

A twofold increase in mean postsynaptic membrane length was associated with ACHE-E6 overexpression (3.8 ± 2.1 μm vs 2.2 ± 1.3 μm). In contrast, PSLs observed in ACHE-I4/E5 embryos (1.9 ± 0.2 μm) were approximately the same as controls. At least 40% of NMJs from uninjected and ACHE-I4/E5 embryos displayed PSLs <3 μm. Such small synapses were not observed in ACHE-E6-injected embryos (Fig. 33). Rather, a class of large (>4 μm) NMJs

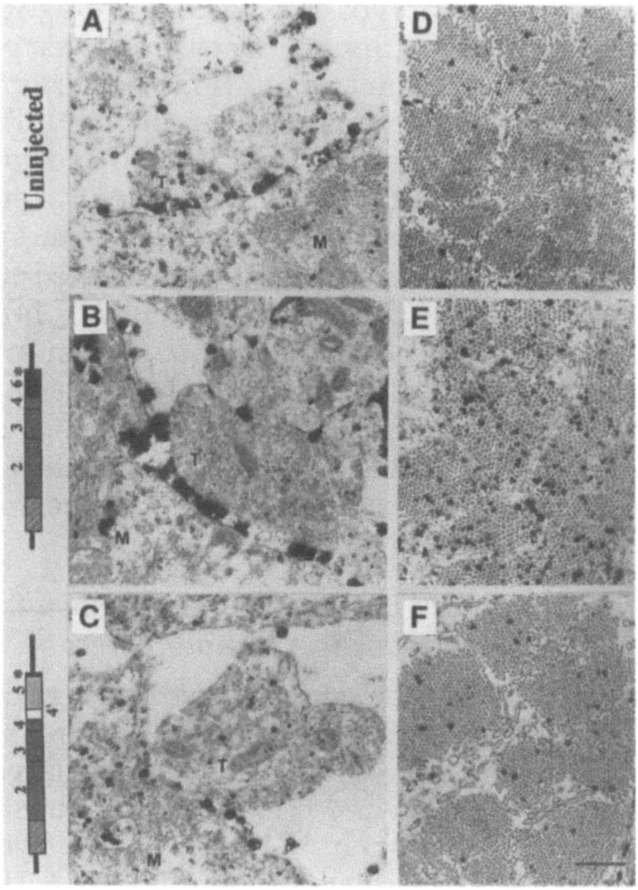

Fig. 32. ACHE-E6 but not I4/E5-derived AChE is localized to NMJs. Two-day-old DNA-injected and control uninjected *Xenopus* embryos raised at 19°C were cytochemically stained for catalytically active AChE and examined by electron microscopy. Note the significantly enhanced staining and increased size displayed by NMJs from ACHE-E6-injected embryos **(B)** as compared to those from control **(A)** and ACHE-I4/E5 injected **(C)** embryos. **(D–F)** Transverse sections of myofibers from control (D), ACHE-E6-injected embryos (E), and ACHE-I4/E5 injected (F) embryos; note the correlation between myotomal and synaptic AChE levels. M, muscle cell; T, nerve terminal. Reprinted from Seidman et al. (1995) with permission from American Society for Microbiology.

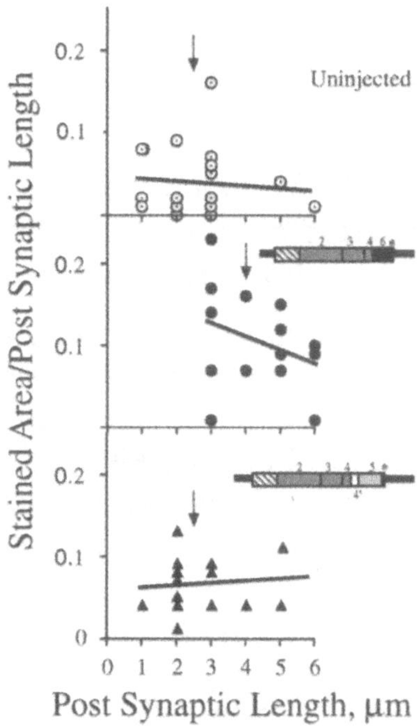

Fig. 33. ACHE-E6, but not ACHE-I4/E5, promotes NMJ maturation. Ratios of AChE-stained area to PSL as a function of length are presented for cross-sections of individual NMJs from embryos injected with ACHE-E6 **(middle)** or ACHE-I4/E5 **(lower)** and raised for 2 d at 21°C as compared with controls **(upper)**. The solid line represents the best-fit line through the data points and the arrow designates the mean PSL for each microinjection group. Note the shift toward larger NMJs in embryos injected with ACHE-E6 compared to those found in ACHE-I4/E5-injected and control embryos. Reprinted from Seidman et al. (1995) with permission from American Society for Microbiology.

rarely observed in control or ACHE-I4/E5-injected embryos became dominant in those transgenic for ACHE-E6. When AChE-stained area was calculated as a function of PSL, ACHE-I4/E5-injected and control uninjected embryos displayed a similar, constant relation between these parameters. In contrast, ACHE-E6-injected embryos displayed a higher ratio for all length categories (Fig. 33). Together, these observations suggested a role for the

E6-derived peptide in localizing AChE to NMJs, and demonstrated that the effects of AChE overexpression on NMJ biogenesis are dependent on localization of the enzyme at the NMJ.

Polarized Accumulation of an Alternative Human Acetylcholinesterase in Epidermal Cells

To study the cellular and subcellular distribution of overexpressed AChE in epidermal cells of embryos injected with ACHE-I4/E5, we examined epidermis from cytochemically stained embryos by electron microscopy. Two types of cells were labeled by AChE reaction product: ciliated epidermal cells (derived from the inner, sensorial, and epidermal layer) and mucous-discharging secretory cells (Billett and Gould, 1971) (Fig. 34A,B). There are many fewer ciliated cells than secretory cells. Yet, the frequency and intensity of staining was higher among the ciliated cells,

---

Fig. 34. *(opposite page)* ACHE-I4/E5-derived AChE accumulates in epidermal cells. Three-day-old DNA-injected and control uninjected *Xenopus* embryos raised at 19°C were fixed, and stained for catalytically active AChE. Electron micrographs shown in **A–D** were taken after extending the staining reaction for 2 h. (A,B) Low magnification image of ciliated (CC) and secretory (SC) epidermal cells from an ACHE-I4/E5-injected (A) or control (B) embryo. Note the massive accumulation of crystalline reaction product at the apex of the ciliated cell from an ACHE-I4/E5-injected embryo. Cilia are not visible at this depth. (C,D) High magnification image of the apical ridge of a ciliated epidermal cell from an ACHE-I4/E5-injected (C), and control, uninjected (D) embryo. Note the appearance of crystals both within vesicles (V) and within the cytoplasm of the injected embryo. **(E,F)** High magnification image of secretory mucous vesicles (MV) from an ACHE-I4/E5-injected (E) and control, uninjected (F) embryo. Note that only secretory vesicles from the ACHE-I4/E5-injected embryo demonstrate accumulation of reaction product in exocytotic vesicles. PG, pigment granule. Reprinted from Seidman et al. (1995) with permission from American Society for Microbiology.

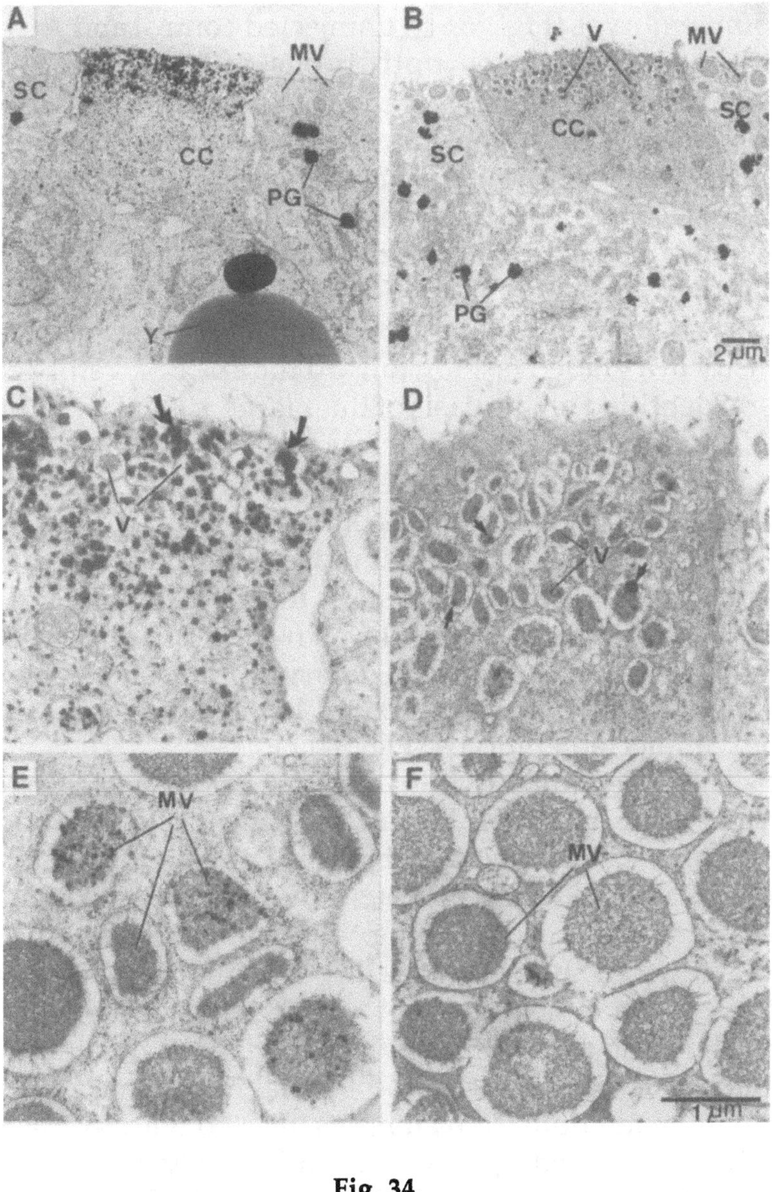

Fig. 34.

with some displaying massive accumulations of reaction product at the apical ridge of the cell (Fig. 34A,C). This irregular mosaic of heavily stained cells corresponded well to the punctuated array of stained cells observed in whole-mount embryos (Fig. 26). In uninjected control and ACHE-E6-injected embryos, both types of epidermal cells displayed scant staining, if any (Fig. 34B,D,F).

Ciliated cells were characterized, in addition to their cilia, by their dense cytoplasm, rich accumulation of mitochondria, and by the presence of numerous small vesicles, most of which were filled with reaction product in stained cells (Fig. 34C). Crystals of reaction product were also observed, however, in the cytoplasm, increasing in size and density in a graded fashion toward the apex. Secretory cells were identified on the basis of their large, distinctive, chondroitin-laden secretory vesicles (Fig. 34E; Nishikawa and Sasaki, 1993). In labeled secretory cells from ACHE-I4/E5-injected embryos, up to 20% of the vesicles were stained positively for AChE. No gross morphological features distinguished stained cells or vesicles in ACHE-I4/E5-injected embryos from unstained ones or from those observed in control or ACHE-E6-injected embryos. Only an occasional crystal of reaction product could be observed in vesicles from uninjected embryos (Fig. 34D,F).

The observation that mature secretory vesicles in epidermis from ACHE-I4/E5-injected embryos stained positively for AChE suggested that this enzyme form may be secreted along with the mucous naturally contained within these vesicles. To determine whether AChE was being excreted from the body, healthy neurula-stage embryos injected with either vector or uninjected, were incubated overnight and the AChE activity found in the medium was compared with that measured in total homogenates. Only incubation medium from embryos injected with ACHE-I4/E5 DNA displayed significant AChE activity, representing up to 40% of the total measured activity (Fig. 35). Together, these observations indicated that some property (or properties) intrinsic to the yet undefined short I4/E5 terminal peptide confer a signal for active epidermal accumulation, polarized subcellular transport, and excretion of AChE.

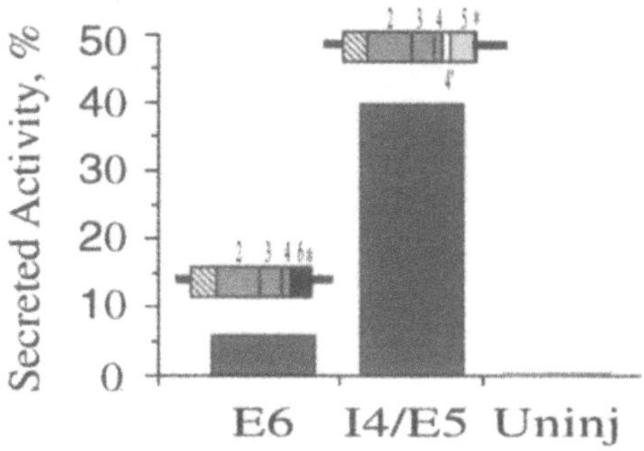

Fig. 35. ACHE-I4/E5, but not ACHE-E6-derived AChE, is excreted. Fractions of total recovered AChE activity (medium + whole cell extract) found in the external growth medium are presented for 20 DNA-injected or control embryos following incubation of neurula stage embryos overnight (d 1 → d 2) in fresh buffer. Note that only embryos expressing ACHE-I4/E5 DNA displayed significant secretion of AChE into the surrounding medium as predicted by the electron microscope studies presented in Fig. 34. Reprinted from Seidman et al. (1995) with permission from American Society for Microbiology.

A Model to Study the Role of Cholinergic Signaling in Neuromuscular Junction Development

The observation that enhanced NMJ development was associated with synaptic AChE overexpression demonstrated that microinjected *Xenopus* embryos could serve as an in vivo experimental model to examine vertebrate synaptogenesis. These findings also suggested that cholinergic signaling may play a role in the formation and/or maintenance of a functional NMJ. Since both AChE and AChR participate in synaptic ACh metabolism, imbalanced expression of either could be expected to affect NMJ formation if, indeed, cholinergic signaling plays a role in this process. Furthermore, if the proteins involved in cholinergic neurotransmission are coordinately regulated, alterations in the expression of one may be expected to feedback on expression of the other. To further study the role of cholinergic neurotransmission in NMJ biogenesis, fertilized *Xenopus* eggs were coinjected with DNA plasmids encoding the four individual subunits of the mouse muscle nAChR (Fig. 15). To reveal changes in NMJ ultrastructure and in synaptic AChE that might be associated with overexpression of the receptor, electron microscopy coupled with cytochemical AChE activity staining was performed.

Overexpression of Mouse Acetylcholine Receptor in Microinjected Xenopus *Embryos*

Microinjection of AChR-encoding DNAs into *Xenopus* oocytes yields increased α-BTX binding and the appearance of electrophysiologically functional receptors on the oocyte membrane (*see* Patrick et al., 1993; *see also Xenopus* Oocyte Microinjection section in Chapter 1). In microinjected embryos, AChR-encoding DNAs induced (over a period of several days) 8–10-fold increased α-BTX binding activity in embryo homogenates (Shapira et al., 1994). A twofold increase in α-BTX binding was also noted in homogenates of ACHE-E6-injected embryos (Shapira et al., 1994). Cross-sections of NMJs from AChR-injected embryos displayed a slightly increased mean postsynaptic membrane length compared to controls, as well as an increase in mean AChE-stained area ($p < 0.05$, Student's *t*-test) (Figs. 36 and 37 and Table 9). Moreover, the ratio of stained area to PSL in NMJs from AChR-injected embryos was approx twofold higher than that observed among controls.

Developmental Implications of Acetylcholine Receptor-Overexpression in Xenopus

Only 10% of the NMJs examined from uninjected embryos developed up to a length of >3.5 μm (Fig. 37A). In contrast, in AChR-injected embryos, approx 30% of the

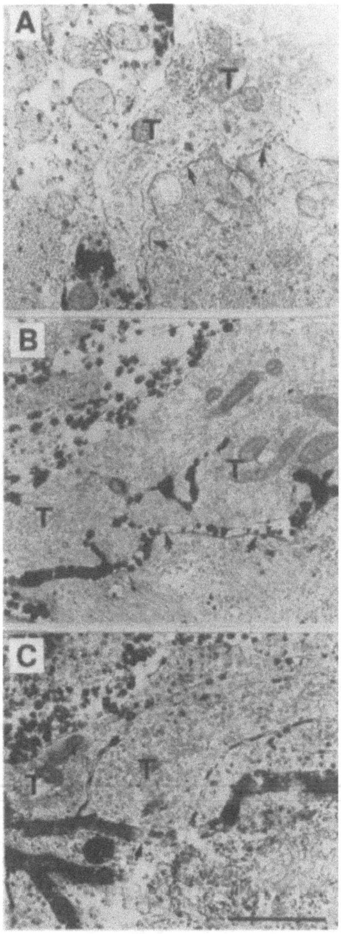

Fig. 36. AChR overexpression and NMJ development. Representative cross-sections of NMJs from 2-d-old *Xenopus* embryos raised at 21°C. **(A)** Control uninjected. **(B)** AChR-injected. **(C)** ACHE-E6-injected. Note the expanded length and width of the NMJs from embryos overexpressing either AChE or AChR, and the enhanced synaptic AChE-staining in both transgenic NMJs. T, nerve terminal. Size bar, 1 μm. Reprinted from Shapira et al. (1994) with permission from National Academy of Sciences.

NMJs were >3.5 μm long. In animals transgenic for human AChE, 50% of the NMJs reached lengths >3.5 μm. Stained area measurements for cross-sections of NMJs in the

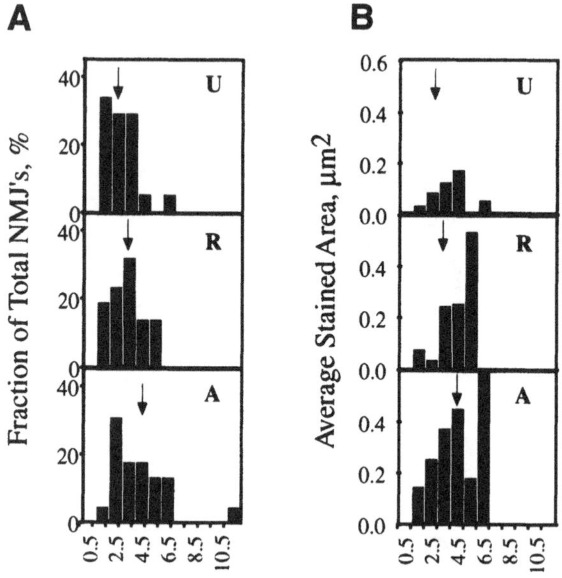

Fig. 37. Morphometric analysis of NMJs from *Xenopus* embryos injected with AChR-encoding DNAs. Micrographs of randomly selected NMJs from 2-d-old embryos microinjected with AChR or ACHE-E6 DNAs and raised at 21°C were subjected to computer-assisted morphometric analysis **(A)** PSL: NMJs were divided into subgroups according to PSL. Figure presents the fraction of the total examined NMJs falling into each group (N = 21–23). Arrows denote weighted mean length for each experimental system. Note the shift toward longer postsynaptic membranes associated with overexpression of both AChR and AChE. **(B)** AChE activity staining: Average AChE-stained area (μm^2) is presented for NMJs in each length subgroup as presented in A. Note the correlation between stained area and length, and the enhanced ratio (area/length) associated with both AChR and AChE overexpression. Reprinted from Shapira et al. (1994) with permission from National Academy of Sciences.

different length subgroups demonstrated a direct relation between stained area and PSL. Nevertheless, within most of the length subgroups, AChE-stained areas were considerably higher in NMJs from AChR-injected embryos than

Table 9
Morphology of NMJs
from *Xenopus* Embryos Expressing AChR or ACHE-E6 DNAs[a]

Group	Length, µm ± Std	Area, µm² ± Std	Area/length, ratio
U (*N* = 21)	2.16 ± 1.31	0.08 ± 0.09	0.037
R (*N* = 22)	2.70 ± 1.22	*0.18 ± 0.15	0.067
A (*N* = 23)	3.80 ± 2.14	0.33 ± 0.29	0.087

[a]Micrographs of the noted numbers (*N*; except * where *N* = 14) of NMJ cross-sections from 2-d-old embryos raised at 21°C were analyzed as detailed in text. Control uninjected embryos (U), or embryos microinjected with AChE- or AChR-encoding DNAs (A and R, respectively) were assessed for postsynaptic membrane length (µm) and AChE-stained cross-sectional area (µm²). Adapted from Shapira et al. (1994).

in controls, although less than that observed following injection of ACHE-E6 (Fig. 37B).

Nerve-Terminal Acetylcholinesterase Expression

In some NMJs, presynaptic AChE staining was noted. The fraction of nerve-labeled NMJs was only 10% (2/21) in control embryos. However, 57% (8/14) of NMJs from AChR-injected and 34% (8/23) of ACHE-E6-injected embryos displayed nerve-terminal labeling (Fig. 38A). Thus, the expression of endogenous *Xenopus* AChE and/or exogenous human AChE was enhanced in some nerve terminals of transgenic embryos. Significantly increased AChE staining was observed in NMJs from ACHE-E6-injected embryos in which nerve terminals were labeled, as compared with nerve-unlabeled NMJs (Fig. 38B). In contrast, there was no difference in AChE-stained area between nerve-stained and nerve-unstained NMJs in control embryos or in those transgenic for nAChR (Fig. 37B). Furthermore, there was no observable correlation between neuronal expression of AChE and PSL (Fig. 38C). Therefore, the increase in AChE-stained area associated with overexpression of ACHE-E6 may reflect widening of synaptic clefts (Fig. 38D,E).

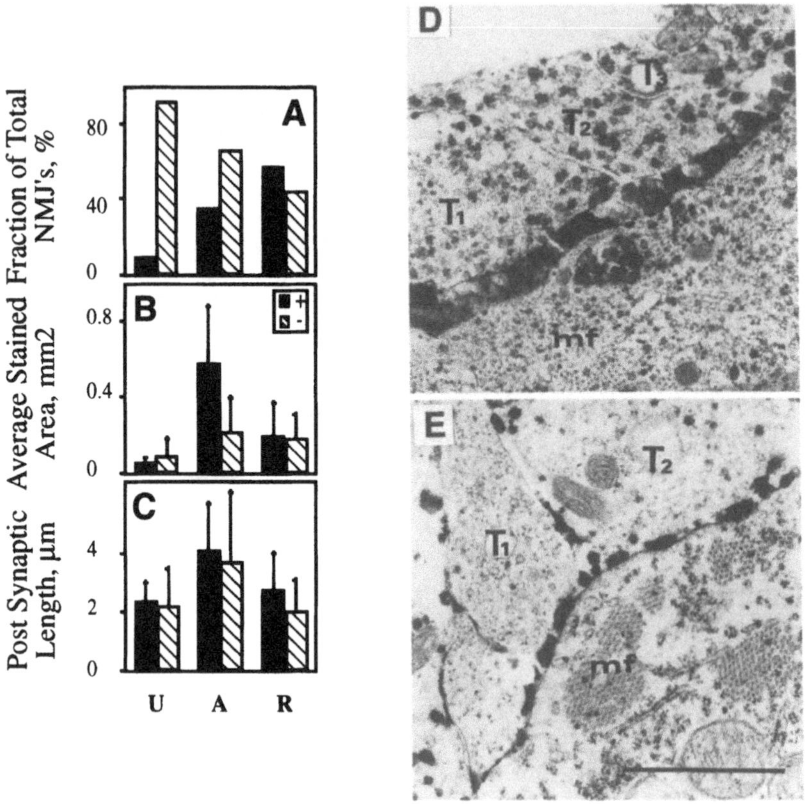

Fig. 38. AChE expression in nerve terminals. **(A)** Enhanced AChE expression in nerve terminals from transgenic embryos: Columns represent fractions of nerve terminals stained (+) or unstained (–) for AChE activity out of total synapses examined. Note the small fraction of AChE-stained nerve terminals in NMJs from uninjected embryos (U) as compared to those from ACHE-E6-injected (A) or AChR-injected (R) embryos. **(B)** Increased synaptic AChE in NMJs displaying nerve-terminal AChE are presented as average AChE-stained area ± SD/NMJ (μm^2). Note that only NMJs from ACHE-E6-injected embryos exhibited a significant correlation between stained area and nerve-terminal AChE activity. **(C)** Nerve-terminal labeling does not affect PSL. Figure presents mean PSL ± SD for NMJs in which the nerve terminal was stained for AChE or not. No significant correlation was observed between nerve-terminal AChE and PSL for any group. **(D,E)** Representative micrographs of NMJs from ACHE-E6-injected embryos: Note the wider synaptic cleft in the NMJ stained for AChE activity in the nerve terminal (D) as compared with the unstained one (E). T, terminal. Reprinted from Shapira et al. (1994) with permission from National Academy of Sciences.

Chapter 4

Conclusions

Over a period spanning some 15 yr, we have exploited microinjected oocytes and embryos of *Xenopus laevis* to clone and characterize the human genes encoding AChE and BuChE and their protein products. Subsequently, we tapped into the enormous versatility of *Xenopus* to continue into the realm of tissue-specific processing and gene function studies. These studies not only revealed crucial elements in the "ChE story" themselves but provided the experimental background for complementary studies in primary cell cultures and transgenic mice. The observation that the morphogenetic implications of AChE overexpression in *Xenopus* NMJs correspond to those observed in NMJs of transgenic mice overexpressing AChE in CNS neurons (Andres et al., in preparation) supports the growing use of microinjected *Xenopus* embryos as a model for vertebrate development, and synaptogenesis in particular.

Characterization of Human Acetylcholinesterase Expressed in *Xenopus*

Native Properties of Heterologous Recombinant Human Acetylcholinesterase

In these studies, we employed microinjected *Xenopus* oocytes and embryos to demonstrate the expression of catalytically active rHAChE from a cloned ACHE coding sequence. Heterologous human AChE, as expressed in *Xenopus*, demonstrated biochemical and immunochemical properties that were indistinguishable from native human AChE. Thus, these experiments verified the cloned ACHE DNA as encoding a true AChE. The fact that we were able to differentiate between rHAChE and endogenous *Xenopus* AChE in complex mixtures using various inhibitors demonstrated that the ligand-binding properties of AChE are intrinsic to its primary amino acid sequence and not dependent on cell-specific posttranslational modifications. These studies also demonstrated the efficacy of the CMV promoter in *Xenopus,* directing the production of 5–10-fold higher levels of heterologous enzyme than those induced by microinjection of in vitro-transcribed mRNA. Together, these observations opened the door for studies into the in vivo regulation and developmental and physiological import of AChE expression using cloned ACHE DNAs to generate transiently transgenic *Xenopus* embryos.

Developmental Implications of Acetylcholinesterase Overexpression

Ectopic overexpression of developmentally important genes in microinjected *Xenopus* embryos often results in gross morphogenic aberrations (McMahon and Moon, 1989; Sokol et al., 1991). Yet, our studies demonstrated that *Xenopus* embryos can tolerate large excesses of catalytically active heterologous AChE without suffering gross morphological or developmental abnormalities. This observation is especially interesting in light of evidence implicating AChE with developmental processes such as gastrulation and cell migration (Drews, 1975; Fitzpatrick and Stent, 1981), nerve outgrowth and differentiation (Layer and Willbold, 1995), and proliferation and differentiation of hematopoietic cells (Lapidot-Lifson et al., 1989,1992; Patinkin et al., 1990; Lev-Lehman et al., 1994; Soreq et al., 1994) (*see* Acetylcholinesterase section in Chapter 1). As neither the overall rate of development nor general morphology of ACHE DNA-injected embryos was altered by 50–100-fold excesses of the active enzyme at the gastrula stage, our findings do not support a role for rHAChE in modulating cell growth, proliferation, or movement in very early *Xenopus* embryogenesis. However, these biological activities may be unassociated with ACh hydrolysis (*see* the next section). Moreover, heterophilic interactions with other molecules may be required. In that case, they may demonstrate more stringent species-specificity and remain undetected in the *Xenopus* system. Thus, a note of caution: It may be necessary to invoke overexpression of autologous *Xenopus* gene(s) for a complete answer in some gene function studies utilizing this expression system.

A Role for Acetylcholinesterase in Cell Adhesion?

Various alternative and noncatalytic roles for AChE have been discussed (Soreq and Zakut, 1993; Layer and Willbold, 1995). In this light, the striking homology between AChE and the *Drosophila* cell adhesion molecule neurotactin might be noted. Neurotactin is a 135-kDa integral membrane glycoprotein located at intercellular junctions in the larval nervous system (de la Escalera et al., 1990). Curiously, the extracellular domain of neurotactin shares high homology with the N-terminal portions of both AChE and BuChE, as well as with thyroglobulin and glutactin. Transfected into nonadhesive S2 cells, neurotactin-encoding DNA promoted heterophilic, but not homophilic, cell adhesion with a subpopulation of cells from gastrula-stage *Drosophila* embryos (Barthalay et al., 1990). Sequence homologies between AChE and the cell surface molecules gliotactin (Auid et al., 1995) and neuroligin (Ichtchenko et al, 1995) have also been noticed. These molecules are known to interact with the neurexin family of neuronal surface receptors (Ullrich et al., 1995). The expression of AChE in migrating cells, its extracellular localization, and its homology to known cell adhesion molecules and ligands therefore predict a role for AChE in processes involving heterophilic cell–cell recognition. This hypothesis might be tested by employing a cell adhesion assay (Barthalay et al., 1990; Hunter et al., 1991) using gastrula-stage cells from microinjected embryos and appropriate acceptor cells of mammalian origin.

Both AChE and neurotactin bear at least one LRE tripeptide—a motor-neuron selective adhesion site present on several other basal lamina proteins including S-laminin, laminin, and agrin (Hunter et al., 1989; Hall, 1995). In addition to promoting cell adhesion, the LRE motif was also shown to inhibit neurite outgrowth of NSC-34 cells in culture (Hunter et al., 1991). NSC-34 cells exhibit several properties characteristic of motor neurons including the expression of ChAT, the synthesis and storage of ACh, and the release of ACh in response to depolarization. Thus, the LRE sequence could, in principle, participate in forming the durable, basal-lamina-resident imprint, which guides motor neurons back to their original synaptic sites following denervation (Anglister and McMahan, 1985). In that case, AChE as a first-stage synaptic protein and a bearer of LRE, could play a role in establishing and maintaining initial synaptic contacts between nerve and muscle (Hall and Sanes, 1993; Dan and Poo, 1994).

ChEs have been implicated in neuron growth and axonal guidance by cytochemical localization (reviewed by Layer and Willbold, 1995). Experimental evidence for a noncatalytic role for AChE in neurite extension and cell adhesion was first reported by Layer et al. (1993). In that study, selective ChE inhibitors were applied to cultured chick tectal cells and assessed for their effects on neurite outgrowth and organization. Whereas the AChE-specific inhibitor BW284C51 (which binds both at the active site and at an additional peripheral site on the protein's surface) completely blocked catalytic activity and neurite extension in a dose-dependent manner, the selective active-site inhibitor echothiophate blocked activity but not neurite growth. BW and at least two specific BuChE inhibitors were also shown to cause defasiculation in several chick-derived cell-culture systems. These findings were subsequently strengthened by the observation that substratum-bound AChE promoted neurite extension of cultured chick sympathetic neurons even after irreversible inhibition of catalytic activity (Small et al., 1995).

Nonassembly of Recombinant Human Acetylcholinesterase in Xenopus

DNA constructs encoding the AChE-T from *Torpedo* (Duval et al., 1992b) and rat (Legay et al., 1993b) give rise to all the globular AChE forms in transfected mammalian cells, including type II amphiphilic monomers. Similarly, human cell lines transfected with ACHE-E6 expressed and secreted homo-oligomers (Velan et al., 1991a; Kronman et al., 1992). In vivo, the predominant secretory AChE form is G4 as observed in FCS, cerebrospinal fluid, neuroblastoma, and adrenal chromaffin cells (Massoulie et al., 1993). Moreover, mammalian cells cotransfected with cDNAs encoding catalytic and noncatalytic AChE subunits (Krejci et al., 1991; Legay et al., 1991a) produced multimeric globular and asymmetric AChEs, indicating that spatial coexistence may normally be the only requirement for multimeric assembly.

AChE-E6 displayed oligomeric assembly in microinjected *Xenopus* oocytes, but not in developing embryos where only monomeric rHAChE was detected. Nonetheless, AChE-E6 was found to accumulate in its natural subcellular compartments within muscle and was correctly transported to the NMJ of transiently transfected tadpoles. These findings are puzzling in light of the demonstration that secretion appears linked to oligomerization in transfected human 293 cells (Velan et al., 1992; Kerem et al., 1993). Furthermore, the lack of demonstrable oligomeric assembly leaves the mode of association of rHAChE with the extracellular surface unexplained. Ramirez et al. (1990) noted, however, that globular AChE monomers and dimers may interact with heparan sulfate proteoglycans—a prominent component of the basal lamina to which the asymmetric AChE forms are probably bound (*see* Acetylcholinesterase section in Chapter 1).

Regulation of Acetylcholinesterase Localization

Muscle Accumulation of Acetylcholinesterase

The native human ACHE promoter includes consensus recognition sites for transcription factors indicative of tissue-specific regulation of transcription (Ben Aziz-Aloya et al., 1993a). However, the CMV promoter used to direct the expression of AChE in *Xenopus* is panactive (Schmidt et al., 1990) and is probably expressed in the embryos in a nonspecific manner relatively early in development. This would explain the high levels of heterologous enzyme observed at the gastrula stage (d 1). Thus, the tissue-specific accumulation of alternative heterologous AChEs likely reflects differential posttranscriptional management of their respective mRNAs or protein products.

Stabilization of AChE mRNA was shown to account for the increased AChE activity accompanying differentiation of cultured myoblasts (Fuentes and Taylor, 1993). The possibility that differential stability of alternative AChE mRNAs plays a role in generating tissue-specific accumulation of rHAChE in *Xenopus* could be resolved using *in situ* hybridization. Tissue-specific accumulation of AChE could also represent posttranslational management of nascent enzyme present in myotomal progenitor cells at the onset of myogenesis. In that case, alternative C-terminal peptides would be implicated in determining the half-life of AChE forms within specific cell types. These results would further imply that cotranslational processes are not required for the correct compartmentalization of AChE in muscle cells. In a similar vein, purified synapsin was shown to be incorporated into nerve terminals in nerve–muscle cocultures prepared from fertilized *Xenopus* eggs microinjected with purified recombinant protein (Lu et al., 1992). In those experiments, functional recruitment of the microinjected molecules was observed as enhanced synaptic performance. Together, the cytochemical data indicate the existence of an intrinsic, evolutionarily conserved property directing the subcellular trafficking of AChE in muscle, and thus explain the accumulation of rHAChE in NMJs of ACHE DNA-injected embryos.

In vivo, AChE is subject to tissue-specific and developmentally regulated posttranslational processing, which gives rise to a complex array of molecular forms varying in their extents of oligomeric assembly, association with noncatalytic subunits, hydrodynamic properties, and sites of subcellular localization (*see* Acetylcholinesterase section in Chapter 1). Transfections of AChE-coding sequences into mammalian cells indicated that alternative splicing could partially account for these multiple molecular forms of AChE (Krejci et al., 1991; Duval et al., 1992b; Legay et al., 1993a,b; Li et al., 1993). Our results in *Xenopus* embryos extend these conclusions by demonstrating that alternative splicing may dictate the final complement of specific AChE catalytic subunits available to particular cell types through selective management of specific AChE mRNAs or polypeptides.

Neuromuscular Junction Localization of Acetylcholinesterase

ACHE mRNA bearing the alternative 3' exon E6 induced specific accumulation of nascent human AChE in muscle and NMJs of transiently transgenic *Xenopus laevis* embryos. Replacement of E6 with an in-frame pseudointronic sequence of similar size directed production of a soluble enzyme species that was neither incorporated into muscle nor NMJs, but was amassed in epidermal cells and excreted into the external culture medium. These observations suggest that the 3'-terminal exon encoding the 40-amino acid C-terminal peptide characteristic of the brain and muscle AChE subtype plays an indispensable role in the NMJ localization of human AChE.

Compartmentalized biosynthesis of AChE (Rotundo 1990; Rossi and Rotundo, 1992; Jasmin et al., 1993), the nAChR (Merlie and Sanes, 1985; Brenner et al., 1990; reviewed by Changeux, 1991), and other NMJ proteins (Pavlath et al., 1989; Ralston and Hall, 1989) suggests that an intricate network of factors are at work to produce and localize NMJ proteins around junctional regions prior to their active accumulation at the NMJ (discussed in the neuromuscular junction sections in Chapter 1). This network could potentially include *cis*-acting elements intrinsic to the mRNA or primary amino acid sequence of NMJ-targeted proteins, as well as *trans*-acting cellular components capable of anchoring or translocating these molecules within the muscle. An example for sequence-dependent translocation and anchoring of a specific mRNA is the developmentally regulated accumulation of Vg-1 in *Xenopus* oocytes (Weeks and Melton, 1987; Yisraeli et al., 1990; Pressman-Schwartz et al., 1992; Kloc and Etkin, 1995). In the case of the NMJ protein β-laminin, a synaptic localization domain was traced to the C-terminal 16-amino acids of the mature protein (Martin et al., 1995).

The present findings indicate that exon E6 and/or its encoded peptide may participate in forming a recognition signal(s) through which yet unidentified cellular elements might mediate accumulation and subcellular localization of AChE. However, search of the GCG data bank did not reveal significant homologies between the E6-encoded peptide and any protein except the related ACh-hydrolyzing enzyme BuChE. It is interesting to note that BuChE—which shares high homology (56%) with AChE in the corresponding C-terminal domain—may also accumulate in NMJs of microinjected *Xenopus* embryos (Loewenstein et al., unpublished observations). Thus, these observations suggest that exon E6 defines a conserved motif for muscle accumulation and trafficking of ChEs, but not necessarily a general signal for targeting proteins to NMJs. Further experiments will be required to determine whether E6 is sufficient to direct unrelated proteins to the NMJ as well.

Stabilization of Acetylcholinesterase in Neuromuscular Junctions

The general state of myotomal overexpression induced by microinjection of ACHE-E6 persists for at least 3 d. The AChE-stained area observed in cross-sections of NMJs from d 3 ACHE-E6-injected embryos was four- to fivefold over that observed in controls. This figure represents half the 10-fold excess measured in d 2 NMJs (Ben Aziz-Aloya et al., 1993a). This apparent reduction in the level of synaptic overexpression from d 2 to d 3 may reflect the overall decline in total rHAChE activity observed during this period. However, since the stained area calculation does not consider the higher density staining observed in NMJs from AChE-E6-injected embryos, it probably represents an underestimate of the actual synaptic AChE content. Therefore, these data may indicate enhanced stability of rHAChE at the NMJ compared to the total pool—a conclusion consistent with the observation that ECM-associated AChE persists *in situ* long after denervation of adult frog skeletal muscle (Anglister and McMahan, 1985). Since the precise relationship between AChE-stained area and the absolute number of AChE molecules is not known, and the linearity of the staining reaction has not been quantitatively determined, it should be useful to apply additional technique(s) to the quantitation of synaptic overexpression (*see*, e.g., Salpeter, 1967; Anglister et al., 1994).

A Novel Secretory Human Acetylcholinesterase?

An mRNA representing the I4 readthrough transcript was reported in mouse tissues, but considered absent in human cells (Li et al., 1993). This difference was attributed to an inherent property of the human ACHE nucleotide sequence (Li et al., 1993). Nonetheless, mRNA carrying the retained I4 sequence was detected in several tumor cell lines of human origin using specific primers and RT-PCR (Karpel et al., 1994). That study did not, however, reveal whether this mRNA species is translated into protein, and what the properties of the putative human readthrough AChE might be. In the present work, RT-PCR indicated the presence of a truncated readthrough mRNA in ACHE-I4/E5-injected *Xenopus* embryos. The protein translated from this short mRNA would lack the cleavable, hydrophobic E5 stretch—precluding GPI-linkage and leaving the relatively hydrophilic I4-encoded peptide. This would explain the high solubility of catalytically active AChE derived from ACHE-I4/E5 DNA. When DNA encoding the mouse readthrough mRNA (characterized by a nonsense codon 40 nucleotides downstream of the E4/I4 boundary) was expressed in COS cells, 97% of the total activity was soluble and secreted into the culture medium (Li et al., 1993). The current study suggests that a truncated human readthrough ACHE mRNA is also translatable, and that it gives rise to a non-muscle, catalytically active, secretory form of AChE in *Xenopus*. Therefore, these observations may offer the first indications for the existence of yet another human AChE subtype representing the common exons E2–E4 and the pseudointron I4.

Since it is not clear what role, if any, AChE naturally plays in early differentiating epidermis, it is yet unclear whether the epidermal accumulation of ACHE-I4/E5-derived AChE represents targeted tissue-specific expression or an anomalous ectopic expression pattern for this AChE in *Xenopus*. Moreover, it is not clear whether transcriptional or posttranscriptional mechanisms were involved in the generation of this short ACHE mRNA species and whether its appearance is unique to this heterologous expression system. Therefore, it is noteworthy that the human readthrough AChE mRNA does carry a sequence (5' ACCTGCCCA 3') that is almost identical (8 of 9 nucleotides) to that contained within a 19-bp consensus destabilizing site that is sufficient to mediate endonucleolytic cleavage of RNA in *Xenopus* oocytes and embryos (Brown et al., 1993). It is therefore possible that our observations further refine this consensus sequence and identify it as an evolutionarily conserved mechanism responsible for 3' shortening of the readthrough mRNA.

The AChE activity generated from ACHE-I4 mRNA is catalytically indistinguishable from the ACHE-E6-derived AChE (Schwarz et al., 1995a) and should be very close in size to the 583 amino acid brain and muscle form. The presence of such a discrete molecule in native human tissues would therefore be difficult to detect and might easily go unnoticed. However, an AChE form with a distinct migration profile in isoelectric focusing was observed in the Alzheimer's brain (Navartnam et al., 1992). Thus, the distinct migration of rHAChE derived from ACHE-I4-mRNA in nondenaturing gel electrophoresis may provide a tool to search for a naturally occurring form of this enzyme species in humans.

A Morphogenic Role for Acetylcholinesterase

Implications of Acetylcholinesterase Overexpression for Synaptogenesis

Overexpression of AChE in developing NMJs was correlated with enhanced NMJ maturation as evaluated primarily by increased postsynaptic membrane lengths. Other parameters indicating accelerated synaptic development included apparent expansion of synaptic clefts and increased secondary folding of postsynaptic membranes. These effects could be attributed to the catalytic activity of the enzyme or to predicted noncatalytic functions of AChE in cell adhesion (*see* A Role for Acetylcholinesterase in Cell Adhesion section in this chapter). Our studies cannot yet distinguish between these possibilities. To resolve this issue, DNA encoding a noncatalytically active mutant of AChE is being expressed in the embryos and evaluated for its effects on NMJ biogenesis.

Nerve-terminal expression of AChE appeared to be one of the features characterizing the large NMJs particular to the transgenic animals. In AChE-E6 injected embryos, larger AChE-stained areas were observed in the synaptic clefts of NMJs labeled presynaptically than in those devoid of nerve-terminal staining. This difference was not observed for NMJs from either control or nAChR-transgenic embryos. In the latter two cases, any detected AChE is necessarily attributable to endogenous *Xenopus* enzyme. Assuming that AChE activity observed in motor nerve terminals represents overexpressed rHAChE, these observations would suggest that neurons may secrete rHAChE into the synaptic cleft and that either muscle or neuron-derived AChE may exert a morphogenic effect on synaptogenesis. These observations are consistent with findings demonstrating neuronal contribution of AChE to the synaptic cleft in adult *Xenopus* NMJs (Anglister, 1991).

Acetylcholine Metabolism and Acetylcholinesterase Overexpression

In animal models, inhibition of AChE in NMJs results in longer mEPCs and increases in intracellular calcium (*see* Acetylcholinesterase section in Chapter 1). Whereas overexpression of AChE in NMJs might be expected to exert the opposite effect (i.e., shortening or interruption of the postsynaptic response), our transiently transgenic *Xenopus* embryos represent the first experimental model for upregulation of synaptic AChE in which it might be possible to test this prediction.

Anglister et al. (1994) have employed a Monte Carlo simulation (Bartol et al., 1991) to test the effects of varying AChE density on miniature end-plate potentials in idealized frog and lizard NMJs. In both vertebrate models, AChE density was noted to be well in excess of the basic minimum required for normal NMJ function—approx 400 mol/μm^2. Nonetheless, AChE densities in the normal adult frog NMJ (approx 600 mol/μm^2) (Anglister et al., 1994) are considerably less than those observed for lizard or mouse (approx 2500 mol/μm^2) (Matthews-Bellinger and Salpeter,

1978) and from the values previously assumed for theoretical treatments of cholinergic signaling in frog NMJs (Land et al., 1984). Under the spatial and kinetic constraints specified in this analysis, a threefold excess of AChE (approximately that observed in our transgenic NMJs) would result in an approx 50% decrease in rise time (t_r) from 100–50 ms, a 20% decrease in fall time (t_f) from 1.5–1.2 ms, and a 30% decrease in the mean number of open channels at the peak (A_c) from 1300–1000. The implications of changes of this magnitude for the developing embryonic NMJ are unclear. However, the directions of the effects are reminiscent of the natural maturation of postsynaptic signals that transpires during synaptogenesis in *Xenopus* and has been attributed to developmentally regulated increases in AChE (Kullberg et al., 1980).

Of course, it may be argued that the adult model bears little resemblance to the embryonic situation in which the absolute and relative densities of AChR and AChE are rapidly changing and for which the necessary kinetic constants have not yet been determined. Furthermore, the geometry exhibited by the immature embryonic NMJ is decidedly distinct from that displayed by an adult frog junction. The actual electrophysiological effect(s) of AChE overexpression should be determined experimentally. Electrophysiological analysis of myotomes from AChE-overexpressing embryos could, in principle, be performed on cultured myotomal cells (Peng et al., 1991) or intact embryos (Kullberg et al., 1977). In addition, parallel studies have been initiated in transgenic mice overexpressing AChE in motoneurons supplying neuronal AChE to the NMJ (Andres et al., in preparation).

A Trophic Role for Acetylcholine?

We observed ultrastructural changes in NMJs of *Xenopus* embryos transiently expressing either human AChE or mouse nAChR. Part of the transgenic NMJs reached sizes unobserved in control embryos, accommodated more AChE within the synaptic cleft, and displayed enhanced AChE staining within nerve terminals. These changes were more conspicuous at 21°C (Shapira et al., 1994; unpublished data) than at 17°C (Ben Aziz-Aloya et al., 1993a; Seidman et al., 1994), as might be expected based on the mammalian origins of the transgenes (Duval et al., 1992b). The fact that both AChE and AChR mediate the physiological effects of ACh suggests a trophic role for ACh in NMJ development (Dan and Poo, 1994). The notion that neurotransmitters may serve a dual role as morphogenic agents has been reviewed (Lauder et al., 1993).

In brain cells, carbamylcholine, an ACh analog, was shown to stimulate DNA replication through muscarinic receptors (Ashkenazi et al., 1989). Experimental evidence of a neurotrophic activity for ACh in motor nerves was derived using cultured *Xenopus* spinal neurons (Zheng et al., 1994). In that study, a gradient of ACh was shown to stimulate positive-turning responses in a calcium-dependent manner. The early onset of ACh sensitivity in developing muscle suggests that cholinergic signaling mechanisms may participate in communicating the approach of the nerve to the muscle (Blackshaw and Warner, 1976). Thus, it would appear that ACh, through its role in stimulating pre- or postsynaptic voltage, current, or ion fluxes, could participate in establishing functional NMJs.

Clinical Implications of Perturbations in Cholinergic Signaling

In humans, ultrastructural and physiological alterations of the NMJ have been associated with congenital AChE and AChR deficiencies (Engel et al., 1977; Wokke et al., 1989; Jennekens et al., 1992; Hutchinson et al., 1993) and may be associated with changes in the balance between these two molecules at the synapse. In some of these syndromes, patients presented, in addition to AChE/AChR deficits, NMJs displaying decreased miniature end-plate potentials, reduced postsynaptic membrane lengths, and severely impaired postsynaptic secondary folding (Smit et al., 1988)—opposite features to those observed in our embryos overexpressing these key cholinergic proteins. It is yet unclear whether the reduced expression of synaptic AChE and/or AChR represents a cause or an outcome of the ultrastructural aberrations observed in these patients. On the other hand, transgenic mice overexpressing AChE in spinal chord motoneurons displayed exaggerated development of the postsynaptic membrane and progressive myopathology (Andres et al., in preparation). Therefore, together, our observations suggest that disturbed regulation of either gene carries ultrastructural consequences for synaptic development. Furthermore, they implicate ACh and balanced cholinergic signaling in the development and maintenance of a functional vertebrate NMJ.

Concluding Remarks

It should be noted that *Xenopus laevis* has made remarkable contributions to our understanding of the vertebrate NMJ—its structure, function, and composition. Until now, *Xenopus* oocyte microinjections have allowed us to clone and characterize the molecules that interlock to create this monumental communications bridge, which has come to represent a paradigm for all synaptic neurotransmission. Now, with the growing popularity and recognized versatility of *Xenopus* embryo microinjections, we can undoubtedly look to *Xenopus* for novel contributions to our understanding of the regulatory processes directing the tissue-specific gene expression and subcellular transport mechanisms, which must necessarily be involved in NMJ biogenesis. Moreover, with the knowledge that manipulations of cholinergic elements induce changes in NMJ development in microinjected *Xenopus* embryos, we can now ally this experimental model with the powerful techniques of modern molecular neurobiology to identify yet unknown genes contributing to the construction and maintenance of a healthy, functional channel for neuromuscular neurotransmission.

Several lines of research represent direct continuations and/or natural outgrowths of the work described here.

1. Electrophysiological determination of the effect(s) of AChE/AChR overexpression on synaptic function;
2. Assessment, using high-speed video microscopy, of the effects of AChE/AChR overexpression on motor function as expressed in swimming behavior of 4–7-d-old tadpoles;
3. Expression of site-directed mutations rendering hAChE catalytically inactive to differentiate between effects that may result from altered ACh metabolism and those involving putative cell adhesion properties of AChE;
4. Microinjection of DNAs encoding nonsynaptic proteins tagged with the E6 coding sequence to test the ability of this C-terminal peptide to direct an irrelevant protein to the NMJ; and
5. Application of advanced molecular biological technologies such as PCR-based differential display to identify and clone genes whose expression may be altered by perturbations in the expression of key cholinergic proteins during NMJ biogenesis.

Thus, the *Xenopus* embryo microinjection system offers openings into a broad spectrum of questions regarding the role and regulation of AChE and other synaptic proteins in the vertebrate NMJ.

Future Directions

At this point in time, much effort is focused on the cloning and characterization of genes active during vertebrate development. These efforts rely heavily on structural and functional homologies with known invertebrate genes and on expression cloning strategies using bioactivity assays in microinjected embryos. In the years to come, however, we can expect to see increasing attention paid to the regulation of those genes at the transcriptional and posttranscriptional levels. These studies will necessarily be preceded by the sometimes laborious and tedious efforts to clone promoters for developmentally important genes. Furthermore, the role of alternative splicing in important nervous system genes represents an area ripe for study in *Xenopus* embryos that has not been exploited. Moreover, a gradual shift toward structure-function studies within individual protein molecules should become increasingly meaningful. Another important trend to watch is the shift toward the study of postgastrulation events. Whereas very early inductive processes still capture the imagination of molecular embryologists, molecular neurobiologists are only beginning to recognize the vast potential in using *Xenopus* embryo microinjections to dissect the events surrounding neuronal differentiation, axon pathfinding, and synapse formation and maturation. In addition, an immense untapped potential of transgenic *Xenopus* methodologies lies in translating voluminous analytical studies currently performed in microinjected *Xenopus* oocytes into the realm of the biologically complex environment of a living organism. Thus, any number of the countless ion channels, receptors, and transporters already characterized in oocytes potentially represent candidates for rewarding studies in embryos. Indeed, site-directed mutagenesis and receptor subunit mixing experiments represent fascinating potential applications of *Xenopus* embryo expression studies. It is our hope that the monumental leap forward achieved in our own research efforts through transient transgenic expression in *Xenopus* embryos will serve as an inspiration to our many colleagues in the field of molecular neurobiology who may have much to gain from this elegant approach to neuroscience research.

Appendices
to Experimental Methodologies

Appendix I

Worldwide *Xenopus* Suppliers

African *Xenopus* Facility

Robert Legg, Owner
PO Box 118 Noordhoek
Republic of South Africa 7985
Tel: (27) 021 891 569
Fax: (27) 021 891 862
$10/frog plus shipping

*Xenopus*1

716 Northside
Ann Arbor, MI 48105
Tel: (313) 426-2083
Fax: (313) 761-6445/(313) 426-7763
Wild-type males and females: $15 each plus shipping
Albino males and females: $20 each plus shipping

Blades Biological

Cowden Edenbridge
Kent, UK
Tel: (44) 01342 850 242
Fax: (44) 01342 850 924
Mature males and females: $5.60–$7.30 each plus shipping

Horst Kahler

Schrammsweg 13b
20249 Hamburg, Germany
Tel: (49) 040 463 109
Fax: (49) 040 463 169
$17.00–$22.50/female (depending on size); $14.50–$16.00/male plus shipping

Please note that before ordering *Xenopus,* local veterinary and environmental authorities should be consulted. *Xenopus laevis* is considered a biologically hazardous species in some locales and may require special import permits (where applicable) and/or housing constraints.

Appendix II

Laboratory Maintenance of *Xenopus laevis* Frogs

Housing

The rearing and maintenance of *Xenopus* in the laboratory has been treated by numerous investigators (Nieuwkoop and Faber 1967,1994; Goldin, 1991; Wu and Gerhart, 1991). In general, the recurrent themes in *Xenopus* husbandry appear to include:

1. Opaque tanks (with tight fitting perforated or mesh lids) housing one frog/3–4 L (1 gal) of water, 20–30 cm deep;
2. A year-round 12-h light/dark cycle; and
3. Water temperature of 17–22°C.

Frogs may be maintained under standing water or continuous flow conditions. Standing water should be changed two to three times/wk, after feeding. Continuous drip systems should result in complete replacement every few days. More sophisticated housing facilities may include continuous circulation of water passed through one or more filter units for removing solid and chemical wastes. Algae accumulating on the sides of the tank need not (and perhaps should not) be removed unless the buildup is extreme. We have maintained frogs under both still water and continuous drip conditions without observing significant differences in the overall health of the animals. Heavily chlorinated water should either be chemically dechlorinated (commercial agents available at your local pet store) or allowed to stand for several days before being added to the tank. Aerating the tanks may also address this problem.

If chloramine is added to the water in your area, it can be removed with a charcoal filter.

Feeding

Several approaches have been taken to feeding *Xenopus* in the laboratory. Although it was traditionally accepted to maintain the frogs exclusively on a meat-based diet of chopped beef heart or liver, the inconvenience and mess have prompted numerous groups to move toward dry foodstuffs. Most *Xenopus* suppliers offer frog chow purported to be nutritionally complete for *Xenopus*, although several investigators report feeding their animals various brands of floating fish pellets. Animals accustomed to meat may take up to 2 wk to resign themselves to dry food; wild frogs maintained in the laboratory may never adjust. Frogs should be fed to satiation two to three times/wk and "leftovers" removed from the tank. Water should be changed 24 h later.

Healing

Xenopus in the laboratory suffer from a variety of health and life-threatening maladies—a few of which have been characterized, many of which have not. The most common disease among laboratory frogs appears to be loosely referred to as "red foot" (bacterial septicemia). Red foot is characterized particularly by hemorrhaging around the belly, thighs, and webbing of the feet (hence "red foot") and edemic swelling of the body. Unless treated, red foot is fatal within one to several days and can rapidly spread between frogs in the tank. We get one or two occurrences of red foot per year, which we treat by adding either 0.5 g/L neomycin sulfate (Sigma) or several drops of mercurochrome (local pharmacy) to the tank for several days (two treatments); this is usually sufficient to stem the tide of infection. However, we do not generally use frogs treated for red foot for at least a month. Remove dead or severely affected animals from the tank.

Nematode infections, characterized by discoloration and sloughing of the skin accompanied by dramatic weight loss have also been reported. The recommended treatment was Ivermectin (PRO-VET, Il; Tel: (800) 435-6902), 0.2 µg/g body wt injected into the dorsal lymph sac, two doses 2 wk apart (Hazel Sive, personal communication) or thiabendazole (Sigma; 100 µg/mL added to the water overnight) (Wu and Gerhart, 1991). Some researchers report adding 100 m*M* NaCl to the water when frogs appear to be suffering from any number of ill-defined infirmities.

Although *Xenopus* frogs can live happily for up to 5 yr or more in the laboratory, periodically a frog, or several frogs (or rarely, an entire tank) will die, during a short period, for no apparent reason. When this happens, we empty the affected tank(s) and clean them with a strong disinfectant cleaning solution such as acid or bleach. However, there are occasions and periods during which nothing seems to help.

Summary

The health and well-being of the female *Xenopus*, in particular, is often the rate-limiting step in performing a successful microinjection experiment. The primary factors affecting our ability to extract good oocytes or eggs from our donor frogs appear to be:

1. The season: Summer seems to be a difficult period for most *Xenopus* labs and a good time to travel;
2. The ambient temperature: Temperatures below 15°C and above 25°C appear to severely depress reproductive virility and frogs may not recover for several months; and
3. The density of frogs in the tank: Although overcrowding may not result in obvious behavioral contraindications, over a period of time, the quality of oocytes and responsiveness to hormone-induced egg-laying will probably decline.

Furthermore, overcrowding, as well as extreme temperatures and inadequate food may increase stress-related ailments such as "red foot" (Hazel Sive, personal communication).

Appendix III

Preparation of Buffers/Reagents for Microinjections

Modified Barth's Medium

		For 5 L
88.0 m*M* NaCl	(5.14 g/L)	25.7 g
1.0 m*M* KCl	(0.075 g/L)	373 mg
0.33 m*M* $Ca(NO_3)_2\ 4H_2O$	(0.078 g/L)	390 mg
0.41 m*M* $CaCl_2$	(0.060 g/L)	300 mg
0.82 m*M* $MgSO_4\ 7H_2O$	(0.099 g/L)	494 mg
2.4 m*M* $NaHCO_3$	(0.2 g/L)	1.0 g
20 m*M* HEPES	(4.77 g/L)	23.8 g

Adjust to pH 7.4 with HCl, autoclave in 500-mL aliquots, and add antibiotics immediately prior to use: 400 mg/L penicillin; 400 mg/L streptomycin.

MMR Buffer

1X MMR	For 1 L 10X
100 m*M* NaCl	58.5 g
2 m*M* KCl	1.49 g
1 m*M* $MgSO_4$	2.46 g
2 m*M* $CaCl_2$	2.94 g
5 m*M* HEPES	11.92 g
1 m*M* EDTA	0.372 g

Adjust pH to 7.8 and autoclave.

2% Cysteine

Prepare fresh in ddH_2O (2 g/100 mL); adjust to pH 7.4–7.6 with 5*N* NaOH. If the solution is cloudy, do not use it; prepare another batch and open a new bottle if necessary.

Human Chorionic Gonadotropin

Sigma cat no. C5297. Prepare a stock of 5000 U/mL ddH_2O and freeze in aliquots of 0.5 mL. For injections, prepare dilutions in an isotonic buffer.

Appendix IV

Microinjection Equipment

1. Surgical kit: The kit should include a small straight edge scissors, fine-tip forceps (Dumont no. 5), hemostat, silk sutures (4/0), and 1/2 circle triangular surgical needles.
2. Dissecting scope: Only a simple scope is required for microinjection, but a more sophisticated apparatus may offer ease of use and facilitate low magnification light or video photography. We have been extremely satisfied with the Zeiss Stemi SV11 stereomicroscope fitted with a MC80 microscope camera.
3. Cold light source: We use an optic fiber lamp.
4. Micromanipulator and capillary holder: Adapt to the selected microinjector.
5. Microinjector: A variety of systems are available for dispensing microinjected materials. For oocyte microinjections, Drummond (Brookmall, PA) offers both manual (model 510X) and electric (Nanoject, model no. 203XV) microinjectors, which dispense relatively precise quantities through the action of a mechanical piston; with both systems, the needles are back-filled with immersion oil, loaded from the tip, and the dispensing pressure is relatively low. The relatively large bore (up to 30 μm) opening of the needle required for this system is tolerated by the oocyte. The more fragile embryos, however, should be injected with a finer tip needle. Therefore, a high-pressure system like the Eppendorf 5242 (Hamburg, Germany) (Fig. 17) is preferred for embryo microinjections. No oil is required and the needles can be preloaded with sample from the back using a microloader tip (Eppendorf) and an appropriate pipetor (Eppendorf Varipette, 10 μL). The Eppendorf 5246 "transjector" offers all the capabilities of the earlier 5242 model, but does not require an external air source. The principal disadvantage of the air-pressure systems is that

the quantity of sample dispensed is harder to control and less reproducible.

6. Microinjection needles: Prepared by pulling capillary tubes. We use Drummond 10-µL microdispenser capillary tubes (cat. no. 210G) for the standard Drummond microdispenser and air-pressure systems. However, the Nanoject automatic injector requires its own special capillaries, which are provided with the unit. For pulling pipets, a relatively simple pipet puller appears to be adequate. We currently use a Kopf (Tujunga, CA) model 720 vertical pipet puller. Opening the tip of the needle may be one of the most difficult steps in achieving consistently good microinjections. Although breaking the tip with a fine-tip forceps is usually satisfactory for oocyte microinjections, this method may be too imprecise for embryo microinjections resulting in microinjection damage. We have had success by simply pricking the tip (under the binocular) against a taut piece of Parafilm.
7. Petri dishes (glass or plastic) 90 × 15 mm for oocyte and egg isolation and 35 × 10 mm for culturing microinjected oocytes or embryos.
8. 15–23°C humidified incubator.

Appendix V

In Vitro Fertilization and Microinjection of *Xenopus* Embryos

1. One day prior to in vitro fertilization, both male (×1) and female (×2–3) frogs are primed with HCG (Sigma cat. no. C5297) in PBS (Fig. 16). Two doses are administered at 5–6-h intervals by injection (2 mL) into the dorsal lymph sac. Females receive 100 U (IU) in the morning followed by 350 IU in the afternoon; male receives 50 IU, then 100 IU. However, note that in an emergency, a single injection of 800–1000 U/female is usually sufficient to induce spawning. Furthermore, while it is apparently essential to prime the male for live mating, we have not found this step limiting for our in vitro fertilizations and have essentially discontinued this practice.

   Animals are left overnight in a bucket of water with a perforated but tightly weighted lid to allow air in but prevent the frogs from escaping. (They are extremely adept at squeezing through small cracks!) Females should begin laying eggs 16–18 h following administration of HCG. However, any eggs laid into the water during the night will be "activated" and therefore unfertilizable. (If you consistently find large quantities of eggs in the water when you arrive in the morning, either reduce the quantity of HCG or inject later in the day.)

   The efficiency of egg-laying is highly variable; all, or some (or none) of the females may lay on any given day. Sometimes spawning may be delayed, eggs being available only in the afternoon. Gentle massage of the lower back and abdomen may stimulate the females to lay and can be employed periodically during the morning if the frogs appear reticent. "Noncooperative" females can be reinjected with 300–400 IU HCG and may lay the next day.

2. Prepare testes by sacrificing male (intradermal injection of 2 mL Tricaine [MS222-Sandoz; 100 mg/mL]) and making a flap in the lower abdomen; the testes lie in the dorsal abdominal region, attached to the base of the fat bodies (long yellowish hose-like structures), and look like miniature off-white kidneys. Dissect out both testes into cold 1X MMR buffer and store on ice. We have found testicular tissue stored in the refrigerator active for several days (up to 1 wk), although the efficiency of fertilization may decline with time.
3. Squeeze out eggs into a small petri dish by spreading the legs of the female between fingers and thumbs and applying firm but gentle and steady pressure on the belly and back, near the hind legs (Fig. 15). Be Patient! Often the eggs will come out in brief spurts, often assisted by the muscle contractions accompanying the frog's attempts to free herself from your grip. If the female is fertile, eggs can be obtained several times during the day, so you need not exhaust the supply all at once. It is a good idea to collect eggs from several frogs, if possible, in the morning and see which ones seem to fertilize best before committing to a single animal for the day's microinjections. We usually perform two or three fertilizations spaced an hour or two apart to maximize efficiency during the day.

   Gently work the eggs into a monolayer with the forceps; no buffer is required at this point—the eggs are protected by their own slimy jelly coat. Note that there is a wide range of natural variation in the coloration of the animal hemisphere of *Xenopus* eggs—from light brown to almost black. This does not appear to reflect the quality or health of the eggs. However, patchy discoloration may reflect poor health and predict low-yield fertilizations.
4. Cut off a small piece of testes, mince lightly with the forceps, and drag around the eggs making contact with all of them. After approx 90 s, add a large volume of 0.1X MMR to dilute sperm and prevent polyspermy; murky buffer at this point is often an indication of poor-health of the eggs and should be viewed with suspicion. Successful fertilization is evidenced by a synchronous rotation of fertilized

eggs, within about 20 min, such that the black animal poles face up. Fertilization, being accompanied by an expansion of the vitelline membrane, provides a type of liquid-filled "sack," allowing the egg free rotation and resulting in a settling of the heavier vegetal pole downward.

5. Remove jelly coat by pouring off MMR buffer and washing eggs with 2% cysteine two times for 3 min each (with swirling); wash immediately (five to six washes) with 0.1X MMR. At this point, viable embryos can be identified as spherically compact cells with slightly darkened coloration at the animal pole. Swollen, more lightly pigmented eggs do not enter the first cleavage cycle. Note: Approximately 60 min postfertilization, an invisible-to-the-eye, internal cortical rotation takes place, establishing the dorsal/ventral axis of the developing embryo. It is important not to disturb the embryos during this period; therefore, try to begin the dejellification as soon as they "turn" (*see* step 4).
6. Injections:
   a. Preincubate embryos in 5% Ficoll (in 0.3X MMR) for 10–15 min prior to injections to cause shrinkage of the swollen vitelline membrane back onto the eggs; this treatment is said to render the embryos more easily penetrable by the microinjection needle, although by our experience it may not be absolutely necessary. For embryo microinjections, we like to open the tip of the injection needle by pricking it against a piece of parafilm, under the dissecting scope. During the hot summer months, we make a "cool" box to keep the embryos at 16–19°C before and immediately after injection by filling a plastic refrigerator storage box with ice and placing a double-layered plastic rack on top of the ice and covering lightly. Embryos can be held at 15°C for several hours without apparent harm to slow the cleavage rate and expand the window of time for microinjection. Lower temperatures, however, may cause irreversible damage.
   b. Begin injections 70–100 min postfertilization. The first cleavage will begin 90–100 min postfertilization depending on the ambient temperature. The cleavage furrow will be clearly visible at the center of the animal pole

once it starts forming. The second cleavage furrow will be initiated approx 1/2 h later, prior to the complete cell division initiated by the first one. We like to inject groups of 5–25 embryos against the inside edge of a square dish after drawing off as much buffer as possible without leaving the embryos exposed to the air. Following microinjections, flush the plate with buffer, collect the embryos, and place them into a small petri dish containing fresh Ficoll/MMR. Note that the strategy for determining when and where to microinject must be established for each experiment. Furthermore, the distribution and fate of microinjected substances must be determined for each experimental protocol.

7. After several hours, remove embryos that appear to have ceased dividing, as they will eventually lyse and may endanger otherwise healthy embryos ("bad neighbor" effect). Transfer remaining embryos to 0.3X MMR without Ficoll through successive dilutions with 0.3X MMR. We find that the more time that elapses between microinjection and this buffer change, the higher the overnight survival rate. Culture embryos overnight at 17–23°C. The rate of development is highly temperature dependent within a range of 15–25°C and may be a factor to consider in experiments involving transient heterologous gene expression.
8. In the morning, remove aborted embryos as soon as possible to avoid the deleterious effects of dead embryos (swelling, absence of defined cell boundaries, and loss of pigment are signs of embryonic death). Score numbers of healthy, dead, and deformed embryos. Replace buffer with fresh 0.3X MMR and transfer embryos gradually into 0.1X MMR or aged tap water during the day. By d 4 or d 5 postfertilization, *Xenopus* embryos have depleted their maternal yolk stores and begin filter feeding. Therefore, at this time we transfer the free swimming tadpoles to an aerated dish and add small quantities of frozen spinach. Methods for raising tadpoles to adulthood have been described in detail (Nieuwkoop and Faber, 1967,1994; Wu and Gerhart, 1991).

Appendix VI

Whole-Mount Staining for Catalytically Active Acetylcholinesterase

(Based on the Original Method Described by Karnovsky and Roots [1964])

1. Stock solutions:

	For 50 mL
a. 0.1*M* Sodium citrate	1.5 g
b. 30 m*M* $CuSO_4$	0.375 g
c. Potassium ferricyanide	0.082 g

   Stock solutions can be stored for several weeks at 4°C.
2. For 10-mL staining solution (mix in the following order with stirring):
   a. 2 mg acetylthiocholine iodide substrate (Sigma cat. no. A5751) in 6.5 mL 0.1*M* acetate buffer, pH 6.0; sodium hydrogen maleate or phosphate buffer can also be used.
   b. 0.5 mL 0.1*M* sodium citrate.
   c. 1.0 mL 30 m*M* $CuSO_4$.
   d. 1.0-mL ddH_2O.
   e. 1.0 mL 5 m*M* Potassium ferricyanide.
      i. The solution should be greenish and transparent, and should be used promptly.
      ii. Final concentrations for whole mount are 0.67 m*M* acetylthiocholine, 5 m*M* sodium citrate, 3 m*M* $CuSO_4$, 0.5 m*M* potassium ferricyanide in 0.1*M* acetate buffer, pH 5.9. Note that the original protocol calls for 1.7 m*M* substrate.
      iii. Prior to staining, embryos can be lightly fixed for 20 min in 4% paraformaldehyde (in 0.6X PBS), rinsed in PBS several times, and equilibrated to pH 6.0 in 0.1*M* acetate buffer.

iv. Incubate embryos in staining solution with very gentle shaking 2–6 h at room temperature or overnight at 4°C. Wash with PBS and refix for 1 h with 2.5% glutaraldehyde. Wash embryos with PBS and transfer to 100% methanol.

v. To render the embryos transparent, clear in Murray's clearing solution (benzyl alcohol:benzyl benzoate 1:2 [BABB]). **Caution!** Remember that this solution is caustic! Wear gloves when handling, use glass staining bottles, and avoid immersing microscope lens in clearing solution. Note that embryos are fragile after clearing and scratch and break easily; handle as infrequently and gently as possible.

Appendix VII

Isolation of *Xenopus* Oocytes Using Collagenase

Instead of separating the oocytes manually, it is possible to remove the connecting tissue surrounding the oocytes with collagenase. Whole sections of ovaries are incubated with shaking at room temperature for 2 h at 19°C in 25 mL Ca^{2+}-deficient Barth's medium (CDB) containing 0.2% type I collagenase. Oocytes are then washed four times in CDB before being transferred to regular Barth's. This treatment has the advantage of minimizing the loss of viable oocytes resulting from mechanical damage and increases the number of oocytes effectively detached from the ovarian tissue. Thus, it is especially useful for isolating large numbers of oocytes. However, collagenase treatment may also cause microscopic damage to the oocyte plasma membrane (Dascal, 1987). Furthermore, it should be noted that collagenase treatment has been observed to depress protein synthesis (Smith et al., 1991), and treated oocytes should be allowed at least 8 h to recover. Therefore, we generally isolate oocytes in the afternoon and perform microinjections the next day.

References

Alder, J., Kanki, H., Valtorta, F., Greengard, P., and Poo, M.-m. (1995) Overexpression of synaptophysin enhances neurotransmitter secretion at *Xenopus* neuromuscular junctions. *J. Neurosci.* **15,** 511–519.

Alder, J., Xie, Z.-P., Valtorta, F., Greengard, P., and Poo, M.-m. (1992) Antibodies to synaptophysin interfere with transmitter secretion at neuromuscular junctions. *Neuron* **9,** 759–768.

Alles, G. A. and Hawes, R. C. (1940) Cholinesterases in the blood of man. *J. Biol. Chem.* **133,** 375–390.

Amaya, E., Musci, T. J., and Kirschner, M. W. (1991) Expression of a dominant negative mutant of the FGF receptor disrupts mesoderm formation in *Xenopus* embryos. *Cell* **66,** 257–270.

Anderson, M. J. and Cohen, M. W. (1977) Nerve-induced and spontaneous redistribution of acetylcholine receptors on cultured muscle cells. *J. Physiol. (Lond.)* **268,** 757–773.

Anderson, M. J., Cohen, M. W., and Zorychta, E. (1977) Effects of innervation on the distribution of acetylcholine receptors on cultured muscle cells. *J. Physiol. (Lond.)* **268,** 731–756.

Anglister, L. (1991) Acetylcholinesterase from the motor nerve terminal accumulates on the synaptic basal lamina of the myofiber. *J. Cell Biol.* **115,** 755–764.

Anglister, L. and McMahan, U. J. (1985) Basal lamina directs acetylcholinesterase accumulation at synaptic sites in regenerating muscle. *J. Cell Biol.* **101,** 735–743.

Anglister, L., Stiles, J. R., and Salpeter, M. M. (1994) Acetylcholinesterase density and turnover number at frog neuromuscular junctions, with modeling of their role in synaptic function. *Neuron* **12,** 783–794.

Apel, E. D. and Merlie, J. P. (1995) Assembly of the post-synaptic apparatus. *Curr. Biol.* **5,** 62–67.

Apel, E. D., Roberds, S. L., Campell, K. P., and Merlie, J. P. (1995) Rapsyn may function as a link between the acetylcholinesterase receptor and the agrin-binding dystrophin-associated glycoprotein complex. *Neuron* **15,** 115–126.

Ashkenazi, A., Ramachandran, J., and Capon, D. J. (1989) Acetylcholine analogue stimulates DNA synthesis in brain-derived cells via specific muscarinic receptor subtypes. *Nature* **340,** 146–150.

Auid, V. J., Fetter, R. D., Broadie, K., and Goodman, C. S. (1995) Gliotactin, a novel transmembrane protein on peripheral glia is required to form the blood-nerve barrier in *Drosophila. Cell* **81,** 757–767.

Baker, L. P. and Peng, H. B. (1993) Tyrosine phosphorylation and acetylcholine receptor cluster formation in cultured *Xenopus* muscle cells. *J. Cell Biol.* **120,** 185–195.

Barnard, E. A. (1988) Separating receptor subtypes from their shadows. *Nature* **335,** 301,302.

Bartels, C. F., Zelinski, T., and Lockridge, O. (1993) Mutation at codon 322 in the human acetylcholinesterase (ACHE) gene accounts for YT blood group polymorphism. *Am. J. Hum. Genet.* **52,** 926–936.

Barthalay, Y., Hipeau-Jacquotte, R., de la Escalera, S., Jimenez, F., and Piovant, M. (1990) *Drosophila* neurotactin mediates heterophilic cell adhesion. *EMBO J.* **9,** 3603–3609.

Bartol, T. M., Jr., Land, B. R., Salpeter, E. E., and Salpeter, M. M. (1991) Monte Carlo simulation of MEPC generation in the vertebrate neuromuscular junction. *Biophys. J.* **59,** 1290–1307.

Beeri, R., Andres, A., Lev-Lehman, E., Timberg, R., Huberman, T., Shani, M., and Soreq, H. (1995) Transgenic expression of human acetylcholinesterase induces progressive cognitive deterioration in mice. *Curr. Biol.* **5,** 1063–1071.

Ben Aziz-Aloya, R. (1994) *Structure and Regulation of the Human Acetylcholinesterase Gene,* PhD Thesis, The Hebrew University of Jerusalem.

Ben Aziz-Aloya, R., Seidman, S., Timberg, R., Sternfeld, M., Zakut, H., and Soreq, H. (1993a) Expression of a human acetylcholinesterase promoter-reporter construct in developing neuromuscular junctions of *Xenopus* embryos. *Proc. Natl. Acad. Sci. USA* **90,** 2471–2475.

Ben Aziz-Aloya, R., Sternfeld, M., and Soreq, H. (1993b) Promoter elements and alternative splicing in the human ACHE gene, in *Progress in Brain Research,* vol. 98 (Cuello, C., ed.), Elsevier, North Holland, pp. 147–153.

Bendig, M. M. (1981) Persistence and expression of histone genes injected into *Xenopus* eggs in early development. *Nature* **292,** 65–67.

Bessereau, J. L., Stratford-Perricaudet, L. D., Piette, J., LePoupon, C., and Changeux, J. P. (1994) In vivo and in vitro analysis of electrical activity-dependent expression of muscle acetylcholine receptor genes using adenovirus. *Proc. Natl. Acad. Sci. USA* **91,** 1304–1308.

Beumer, T. L., Veenstra, G. J. C., Hage, W. J., and Destree, O. H. J. (1995) Whole-mount immunocytochemistry on *Xenopus* embryos using far-red fluorescent dyes. *TIG* **11,** 9.

Bhushan, A., Lin, H. Y., Lodish, H. F., and Kinter, C. R. (1994) The transforming growth factor β type II receptor can replace the activin type II receptor in inducing mesoderm. *Mol. Cell. Biol.* **14,** 4280–4285.

Billet, F. S. and Gould, R. P. (1971) Fine structural changes in the differentiating epidermis of *Xenopus laevis* embryos. *J. Anat.* **108,** 465–480.

Blackshaw, S. and Warner, A. (1976) Onset of acetylcholine sensitivity and end-plate activity in developing myotome muscles of *Xenopus. Nature (Lond.)* **262,** 217,218.

Blitz, I. L. and Cho, K. W. Y. (1995) Anterior neuroectoderm is progressively induced during gastrulation: the role of the *Xenopus* homeobox gene *orthodenticle. Development* **121,** 993–1004.

Bon, S., Rosenberry, T. L., and Massoulie, J. (1991) Amphiphilic, glycophosphatidylinositol-specific phospholipase-C (PI-PLC) insensitive monomers and dimers of acetylcholinesterase. *Cell. Mol. Neurobiol.* **11,** 157–172.

Bonham, J. R., Dale, G., and Atack, J. R. (1987) Neural tube defect-specific acetylcholinesterase: its properties and quantitation in the detection of anencephaly and spina bifida. *Clin. Chim. Acta* **170,** 69–77.

Boschetti, N., Liao, J., and Brodbeck, U. (1994) The membrane form of acetylcholinesterase from rat brain contains a 20 kDa hydrophobic anchor. *Neurochem. Res.* **19,** 359–365.

Brandan, E., Maldonado, M., Garrido, J., and Inestrosa, N. C. (1985) Anchorage of collagen-tailed acetylcholinesterase to the extracellular matrix is mediated by heparan sulfate proteoglycans. *J. Cell Biol.* **101,** 985–992.

Brenner, H. R., Witzmann, V., and Sakmann, B. K. (1990) Imprinting of acetylcholine receptor messenger RNA accumulation in mammalian neuromuscular junctions. *Nature* **344,** 544–547.

Brown, L. M., Blair, A., Gibson, R., Everett, G. D., Cantor, K. P., Shuman, L. M., Burmeister, L. F., Van Lier, S. F. and Dick, F. (1990) Pesticide exposures and other agricultural risk factors for leukemia among men in Iowa and Minnesota. *Cancer Res.* **50,** 6585–6591.

Brown, B. D., Zipkin, I. D., and Harland, R. M. (1993) Sequence-specific endonucleolytic cleavage and protection of mRNA in *Xenopus* and *Drosophila. Genes Dev.* **7,** 1620–1631.

Buananno, A. and Merlie, J. P. (1986) Transcriptional regulation of nicotinic acetylcholine receptor genes during muscle development. *J. Biol. Chem.* **261,** 11,452–11,455.

Buchanan, J. A., Sun, Y.-A., and Poo, M.-m. (1989) Studies of nerve-muscle interactions in *Xenopus* cell culture: fine structure of early functional contacts. *J. Neurosci.* **9,** 1540–1554.

Burstein, S. A., Adamson, J. W., and Harker, L. A. (1980) Megakaryocytopoiesis in culture: modulation by cholinergic mechanisms. *J. Cell. Physiol.* **54,** 201–208.

Burstein, S. A., Boyd, C. N., and Dale, G. L. (1985) Quantitation of megakaryocytopoiesis in liquid culture by enzymatic determination of acetylcholinesterase. *J. Cell. Physiol.* **122,** 159–165.

Burstein, S. A. and Harker, L. A. (1983) Control of platelet production. *Clin. Haematol.* **12,** 3–22.

Cahine, K. G., Walker, W., and Goldman, D. (1992) A 102 base pair sequence of the nicotinic acetylcholine receptor delta-subunit gene confers regulation of muscle electrical activity. *Development* **115,** 212–219.

Campanelli, J. T., Roberds, S. L., Campbell, K. P., and Scheller, R. H. (1994) A role for dystrophin-associated glycoproteins and utrophin in agrin-induced AChR clustering. *Cell* **77,** 663–674.

Carbonetto, S. and Lindenbaum, M. (1995) The basement membrane at the neuromuscular junction: a synaptic mediatrix. *Curr. Opin. Neurobiol.* **5,** 596–605.

Carraway, K. L., III and Burden, S. J. (1995) Neuregulins and their receptors. *Curr. Opin. Neurobiol.* **5,** 605–612.

Cary, R. B. and Klymkowsky, M. W. (1995) Disruption of intermediate filament organization leads to structural defects at the intersomite junction in *Xenopus* myotomal muscle. *Development* **121,** 1041–1052.

Changeux, J. P. (1991) Compartmentalized transcription of acetylcholine receptor genes during motor endplate epigenesis. *New Biol.* **3,** 413–429.

Chao, L. P., Kan, K. K., Angelini, C., and Keesey, J. (1982) Autoimmune neuromuscular disease induced by a preparation of choline acetyltransferase. *Exp. Neurol.* **75,** 23–35.

Christian, J. L. and Moon, R. T. (1993) Interactions between Xwnt-8 and Spemann organizer signaling pathways generate dorsoventral patterns in the embryonic mesoderm of *Xenopus*. *Genes Dev.* **7,** 13–28.

Chuang, H. Y. K. (1973) Cholinesterase activities of intact platelets as measured by an improved radiometric method. *Biochim. Biophys. Acta* **321,** 546–552.

Cohen, M. W. (1980) Development of an amphibian neuromuscular junction *in vivo* and in culture. *J. Exp. Biol.* **89,** 43–56.

Cornell, R. A. and Kimmelman, D. (1994) Activin mediated mesoderm induction requires FGF. *Development* **120,** 453–462.

Coyle, J. T., Price, D. L., and DeLong, M. R. (1983) Alzheimer's disease: a disorder of cortical cholinergic innervation. *Science* **219,** 1184–1189.

Cunliffe, V. and Smith, J. C. (1994) Specification of mesodermal pattern in *Xenopus laevis* by interactions between Brachyury, noggin, and Xwnt-8. *EMBO J.* **13,** 349–359.

Dale, H. H. (1914) The action of certain esters and ethers of choline and their relation to muscarine. *J. Pharmacol. Exp. Ther.* **6,** 147–190.

Dale, L. and Slack, J. M. W. (1987) Fate map for the 32-cell stage of *Xenopus laevis*. *Development* **99,** 527–551.

Dan, Y. and Poo, M.-m. (1994) Retrograde interactions during formation and elimination of neuromuscular synapses. *Curr. Biol.* **4,** 95–100.

Dascal, N. (1987) The use of *Xenopus* oocytes for the study of ion channels. *CRC Crit. Rev. Biochem.* **22,** 317–387.

Davis, R. E., Emmerling, M. R., Jaen, J. C., Moos, W. H., and Spiegel, K. (1993) Therapeutic intervention in dementia. *Crit. Rev. Neurobiol.* **7,** 41–83.

DeChiara, T. M., Bowen, D. C., Balenzuela, D. M., Simmons, M. V., Poueymirou, W. T., Tomas, S., Kinetz, E., Compton, J. D. L., Rojas, E., Park, J. S., Smith, C., Distefano, P. S., Glass, D. J., Burden, S. J., and Yancopoulos, G. D. (1996) The receptor tyrosine kinase MuSK is required for neuromuscular junction formation in vivo. *Cell* **85,** 501–512.

de la Escalera, S., Bockamp, E. O., Moya, F., Piovant, M., and Jimenez, F. (1990) Characterization and gene cloning of neurotactin, a *Drosophila* transmembrane protein related to cholinesterases. *EMBO J.* **9,** 3593–3601.

Dent, J. A., Poulson, A. G., and Klymkowsky, M. W. (1989) A whole-mount immunocytochemical analysis of the expression of the intermediate filament protein vimentin in *Xenopus*. *Development* **105,** 61–74.

Deuchar, E. M. (1966) Xenopus: *The South African Clawed Frog,* Methuen, London.

Deuchar, E. M. (1975) *Biochemical Aspects of Amphibian Development,* Methuen, London.

Ding, D. and Lipshitz, H. D. (1993) Localized mRNAs and their functions. *Bioassays* **15,** 651–658.

Drews, E. (1975) Cholinesterase in embryonic development. *Prog. Histochem. Cytochem.* **7,** 1–52.

Dreyfus, P. A., Rieger, F., and Pincon-Raymon, M. (1983) Acetylcholinesterase of mammalian neuromuscular junctions: presence of tailed asymmetric acetylcholinesterase in synaptic basal lamina and sarcolemma. *Proc. Natl. Acad. Sci. USA* **80,** 6698–7702.

Dreyfus, P. A., Seidman, S., Pincon-Raymond, M., Murawsky, M., Schejter, E., Zakut, H., and Soreq, H. (1989) Tissue-specific processing and polarized compartmentalization of clone-produced cholinesterase in microinjected *Xenopus* oocytes. *Cell. Mol. Neurobiol.* **9,** 323–341.

Dudai, Y., Herzberg, M., and Silman, I. (1973) Molecular structures of acetylcholinesterase from electric organ tissue of the electric eel. *Proc. Natl. Acad. Sci. USA* **70,** 2473–2476.

Dumont, J. N. (1972) Oogenesis in *Xenopus laevis* (Daudin) 1. Stages of oocyte development in laboratory maintained animals. *J. Morphol.* **136,** 153–180.

Duval, N., Krejci, E., Grassi, J., Coussen, F., Massoulie, J., and Bon, S. (1992a) Molecular architecture of AChE collagen tailed forms: construction of a glycolipid tailed tetramer. *EMBO J.* **11,** 3255–3261.

Duval, N., Massoulie, J., and Bon, S. (1992b) H and T subunits of acetylcholinesterase from *Torpedo,* expressed in COS cells, generate all types of globular forms. *J. Cell Biol.* **118,** 641–653.

Eckstein, F. (1985) Nucleoside phosphorothioates. *Annu. Rev. Biochem.* **54,** 367–402.

Ehrlich, G., Viegas-Pequignot, E., Ginzberg, D., Sindel, L., Soreq, H., and Zakut, H. (1992) Mapping the human acetylcholinesterase gene to chromosome 7q22 by fluorescent *in situ* hybridization coupled with selective PCR amplification from a somatic hybrid cell panel and chromosome-sorted DNA libraries. *Genomics* **13,** 1192–1197.

Ellman, G. L., Courtney, D. K., Anders, V., and Featherstone, R. M. (1961) A new and rapid colorimetric determination of acetylcholinesterase activity. *Biochem. Pharmacol.* **7,** 88–95.

Engel, A. G., Lambert, E. H., and Gomez, M. R. (1977) A new myasthenic syndrome with endplate acetylcholinesterase deficiency, small nerve terminals and reduced acetylcholine release. *Ann. Neurol.* **1,** 315–330.

Engel, A. G., Lambert, E. H., and Santa, T. (1973) Study of long-term anticholinesterase therapy. *Neurology* **23,** 1273–1281.

Etkin, L. D., Pearman, B., Roberts, M., and Bektesh, S. L. (1984) Replication, integration, and expression of exogenous DNA injected into fertilized eggs of *Xenopus laevis. Differentiation* **26,** 194–202.

Evers, J., Laser, M., Sun, Y.-A., Xie, Z.-P., and Poo, M.-m. (1989) Studies of nerve-muscle interactions in *Xenopus* cell culture: analysis of early synaptic currents. *J. Neurosci.* **9,** 1523–1539.

Eyal-Giladi, H. (1954) Dynamic aspects of neural induction. *Arch. Biol.* **65,** 180–259.

Fakuda, K., Kubo, T., Akiba, I., Maeda, A., Mishina, M., and Numa, S. (1987) Molecular distinction between muscarinic acetylcholine receptor subtypes. *Nature* **327,** 623–625.

Ferreiro, B., Kintner, C., Zimmerman, K., Anderson, D., and Harris, W. A. (1994) XASH genes promote neurogenesis in *Xenopus* embryos. *Development* **120,** 3649–3655.

Fischer, M., Ittah, A., Liefer, I., and Gorecki, M. (1993) Expression and reconstitution of biologically active human acetylcholinesterase from *Escherichia coli. Cell. Mol. Neurobiol.* **13,** 25–38.

Fitzpatrick-McElligot, S. and Stent, G. S. (1981) Appearance and localization of acetylcholinesterase in embryos of the leech *Helodbella triserialis. J. Neurosci.* **1,** 901–907.

Flucher, B. E. and Daniels, M. D. (1989) Distribution of Na^+ channels and ankyrin in neuromuscular junctions is complementary to that of acetylcholine receptors and the 43 KD protein. *Neuron* **3,** 163–175.

Fournier, D., Bride, J.-M., Hoffman, F., and Karch, F. (1992a) Acetylcholinesterase: two types of modifications confer resistance to insecticide. *J. Biol. Chem.* **267,** 14,270–14,274.

Fournier, D., Mutero, A., and Rungger, D. (1992b) Drosophila acetylcholinesterase: expression of a functional precursor in *Xenopus* oocytes. *Eur. J. Biochem.* **203,** 513–519.

Frank, D. and Harland, R. M. (1991) Transient expression of XMyoD in non-somitic mesoderm of *Xenopus* gastrulae. *Development* **113,** 1387–1393.

Froehner, S. C. (1991) The submembrane machinery for nicotinic acetylcholine receptor clustering. *J. Cell Biol.* **114,** 1–7.

Froehner, S. C., Luetje, C. W., Scotland, P. B., and Patrick, J. (1990) The postsynaptic 43K protein clusters muscle nicotinic acetylcholine receptors in *Xenopus* oocytes. *Neuron* **5,** 403–410.

Fuentes, M. E., Rosenberry, T. L., and Inestrosa, N. C. (1988) A 13 kDa fragment is responsible for the hydrophobic aggregation of brain G4 acetylcholinesterase. *Biochem. J.* **256,** 1047–1050.

Fuentes, M. E. and Taylor, P. (1993) Control of acetylcholinesterase gene expression during myogenesis. *Neuron* **10,** 679–687.

Gautam, M., Noakes, P. G., Moscoso, L., Rupp, F., Scheller, R. H., Merlie, J. P., and Sanes, J. R. (1996) Defective neuromuscular synaptogenesis in agrin-deficient mice. *Cell* **85,** 525–535.

Gautam, M., Noakes, P. G., Mudd, J., Nichol, M., Chu, G. C., Sanes, J. R., and Merlie, J. P. (1995) Failure of postsynaptic specialization to develop at neuromuscular junctions of rapsyn-deficient mice. *Nature* **377,** 232–236.

Gee, S. H., Montanaro, F., Lindenbaum, M. H., and Carbonetto, S. (1994) Dystroglycan-α, a dystrophin-associated glycoprotein, is a functional agrin receptor. *Cell* **77,** 675–686.

Gennari, K. and Brodbeck, U. (1985) Molecular forms of acetylcholinesterase from human caudate nucleus: comparison of salt-soluble and detergent-soluble tetrameric enzyme species. *J. Neurochem.* **44,** 697–704.

Gennari, K., Brunner, J., and Brodbeck, U. (1987) Tetrameric detergent-soluble acetylcholinesterase from human caudate nucleus: subunit composition and number of active sites. *J. Neurochem.* **49,** 12–18.

Gibney, G. and Taylor, P. (1990) Biosynthesis of *Torpedo* acetylcholinesterase in mammalian cells. Functional expression and mutagenesis of the glycophospholipid-anchored form (published erratum appears in *J. Biol. Chem.* **265[32],** 20,051). *J. Biol. Chem.* **265,** 12,576–12,583.

Giebelhaus, D. H., Eib, D. W., and Moon, R. T. (1988) Antisense RNA inhibits expression of membrane skeleton protein 4.1 during embryonic development of *Xenopus. Cell* **53,** 601–615.

Gilbert, S. (1994) *Developmental Biology* (4th ed.), Sinauer, Sunderland.

Gilson, M. K., Straatsma, T. P., McCammon, J. A., Ripoll, D. R., Faerman, C. H., Axelsen, P. H., Silman, I., and Sussman, J. L. (1994) Open "back door" in a molecular dynamics simulation of acetylcholinesterase. *Science* **263,** 1276–1278.

Gimlich, R. L. (1991) Fluorescent dextran clonal markers, in *Methods in Cell Biology,* vol. 36 (Kay, B. K. and Peng, H. B., eds.), Academic, San Diego, pp. 285–297.

Gindi, T. and Knowland, J. (1979) The activity of cholinesterases during the development of *Xenopus laevis. J. Embryol. Exp. Morphol.* **51,** 209–215.

Glass, D. J., Bowen, D. C., Stitt, T. N., Radziejewski, C., Bruno, J., Ryan, T. E., Gies, D. R., Shah, S., Matsson, K., Burden, S. J., DiStefano, P. S., Valenzuela, D. M., DeChiara, T. M., and Yancopoulos, G. D. (1996) Agrin acts via a MuSK receptor complex. *Cell* 85, 513–523.

Glazer, E. J., Baker, T., and Riker, W. F. (1978) The neuropathology of DFP at cat soleus neuromuscular junctions. *J. Neurocytol.* **7,** 741–758.

Gnatt, A., Loewenstein, Y., Yaron, A., Schwarz, M., and Soreq, H. (1994) Site-directed mutagenesis of active site residues reveals plasticity of human butyrylcholinesterase in substrate and inhibitor interactions. *J. Neurochem.* **62,** 749–755.

Godfrey, E. W., Nitkin, R. M., Wallace, B. G., Rubin, L. L., and McMahan, U. J. (1984) Components of *Torpedo* electric organ and muscle that cause aggregation of acetylcholine receptors on cultured muscle cells. *J. Cell Biol.* **99,** 615–627.

Godsave, S., Dekker, E.-J., Holling, T., Tjadine, H., Pannese, M., Boncinelli, E., and Durston, A. (1994) Expression patterns of Hoxb genes in the *Xenopus* embryo suggest roles in anteroposterior specification of the hindbrain and in dorsoventral patterning of the mesoderm. *Dev. Biol.* **166,** 465–476.

Goldin, A. (1991) Maintenance of *Xenopus laevis* and oocyte injection, in *Methods in Enzymology,* vol. 207 (Rudy, B. and Iverson, L. E., eds.), Academic, New York, pp. 266–279.

Graff, J. M., Thies, R. S., Song, J. J., Celeste, A. J., and Melton, D. A. (1994) Studies with a *Xenopus* BMP receptor suggest that ventral mesoderm-inducing signals override dorsal signals *in vivo*. *Cell* **79,** 169–179.

Greenfield, S. (1984) Acetylcholinesterase may have novel functions in the brain. *Trends Neurosci.* **7,** 364–368.

Gundersen, C. B. and Miledi, R. (1983) Acetylcholinesterase activity of *Xenopus laevis* oocytes. *Neuroscience* **10,** 1487–1495.

Gurdon, J. B., Lane, C. D., Woodland, H. R., and Marbaix, G. (1971) Use of frog eggs and oocytes for the study of mRNA and its translation in living cells. *Nature* **233,** 177–182.

Gurdon, J. B., Mohun, T. J., Taylor, M. V., and Sharpe, C. R. (1989) Embryonic induction and muscle gene activation. *Trends Genet.* **5,** 51–56.

Hall, Z. W. (1973) Multiple forms of acetylcholinesterase and their distribution in endplate and non-endplate regions of rat diaphragm muscle. *J. Neurobiol.* **4,** 343–361.

Hall, Z. W. (1995) Laminin β2 (S-laminin): a new player at the synapse. *Nature* **269,** 362,363.

Hall, Z. W. and Sanes, J. R. (1993) Synaptic structure and development: the neuromuscular junction. *Cell* **72**/*Neuron* **10,** 99–121.

Hall, L. M. and Spierer, P. (1986) The *Ace* locus of *Drosophila melanogaster*: structural gene for acetylcholinesterase with an unusual 5' leader. *EMBO J.* **5,** 2949–2954.

Hamilton, L. (1969) The formation of somites in *Xenopus*. *J. Embryol. Exp. Morphol.* **22,** 253–264.

Harland, R. M. (1991) *In situ* hybridization: an improved whole mount method for *Xenopus* embryos, in *Methods in Cell Biology*, vol. 36 (Kay, B. K. and Peng, H. J., eds.), Academic, San Diego, pp. 685–695.

Harland, R. M. and Misher, L. (1988) Stability of RNA in developing *Xenopus* embryos and identification of a destabilizing sequence in TFIII RNA. *Development* **102,** 837–852.

Harvey, R. P. (1990) The *Xenopus* MyoD1 gene: an unlocalized maternal mRNA predates lineage-restricted expression in the early embryo. *Development* **108,** 669–680.

Hausen, P. and Riebesell, M. (1991) *The Early Development of* Xenopus Laevis: *An Atlas of the Histology*. Springer-Verlag, Berlin.

He, X., Saint-Jeannet, J.-P., Woodgett, J. R., Varmus, H. E., and Dawid, I. B. (1995) Glycogen synthetase kinase-3 and dorsal-ventral patterning in *Xenopus* embryos. *Nature* **374,** 617–622.

Heasman, J., Ginsberg, D., Geiger, B., Goldstone, K., Pratt, T., Yoshida-Noro, C., and Wylie, C. (1994) A functional test for maternally inherited cadherin in *Xenopus* shows its importance in cell adhesion at the blastula stage. *Development* **120,** 49–57.

Heasman, J., Holwill, S., and Wylie, C. C. (1991) Fertilization of cultured *Xenopus* oocytes and use in studies of maternally inherited molecules. *Methods Cell Biol.* **36,** 213–235.

Hemmati-Brivanlou, A., Frank, D., Bolce, M., Brown, B. D., Sive, H. L., and Harland, R. M. (1990) Localization of specific mRNAs in *Xenopus* embryos. *Development* **106,** 611–617.

Hemmati-Brivanlou, A., Kelly, O. G., and Melton, D. A. (1994) Follistatin, an antagonist of activin is expressed in the Spemann organizer and displays direct neuralizing activity. *Cell* **77,** 283–296.

Hemmati-Brivanlou, A. and Melton, D. A. (1994) Inhibition of activin receptor signaling promotes neuralization in *Xenopus. Cell* **77,** 273–281.

Holt, C. E., Lemaire, P., and Gurdon, J. B. (1994) Cadherin-mediated cell interactions are necessary for the activation of MyoD in *Xenopus* mesoderm. *Proc. Natl. Acad. Sci. USA* **91,** 10,844–10,848.

Hopwood, N. D. and Gurdon, J. B. (1990) Activation of muscle genes without myogenesis by ectopic expression of MyoD in frog embryo cells. *Nature* **347,** 197–200.

Hopwood, N. D., Pluck, A., and Gurdon, J. B. (1991) *Xenopus* Myf-5 marks early muscle cells and can activate muscle genes ectopically in early embryos. *Development* **111,** 551–560.

Hunter, D. D., Cashman, N., Morris-Valero, R., Bullock, J. W., Adams, S. P., and Sanes, J. R. (1991) An LRE (Leucine-Arginine-Glutamate)-dependent mechanism for adhesion of neurons to S-laminin. *J. Neurosci.* **11,** 3960–3971.

Hunter, D. D., Parker, B. E., Bullock, J. W., Adams, S. P., Merlie, J. P., and Sanes, J. R. (1989) Primary sequence of a motor-neuron selective adhesive site in the synaptic basal lamina protein S-laminin. *Cell* **59,** 905–913.

Hutchinson, D. O., Walls, T. J., Nakano, S., Camp, S., Taylor, P., Harper, C. M., Groover, R. V., Peterson, H. A., Jamieson, D. G., and Engel, A. G. (1993) Congenital endplate acetylcholinesterase deficiency. *Brain* **116,** 633–653.

Ichtchenko, K., Hata, Y., Nguyen, T., Ullrich, B., Missler, M., Moomaw, C., and Sudhof, T. C. (1995) Neuroligin 1: a splice site-specific ligand for β-neurexins. *Cell* **81,** 435–443.

Inestrosa, N. C. and Perelman, A. (1989) Distribution and anchoring of molecular forms of acetylcholinesterase. *Trends Pharmacol. Sci.* **10,** 325–329.

Inestrosa, N. C., Roberts, W. L., Marshall, T. L., and Rosenberry, T. L. (1987) Acetylcholinesterase from bovine caudate nucleus is attached to membranes by a novel subunit distinct from those of acetylcholinesterases in other tissues. *J. Biol. Chem.* **262,** 4441–4444.

Jasmin, B. J., Lee, R. K., and Rotundo, R. L. (1993) Compartmentalization of acetylcholinesterase RNA and enzyme at the vertebrate neuromuscular junction. *Neuron* **11,** 467–477.

Jennekens, F. G. I., Hesselmans, L. F. G. M., Veldman, H., Jansen, E. N. H., Spaans, F., and Molenaar, P. C. (1992) Deficiency of acetylcholine receptors in a case of endplate acetylcholinesterase deficiency: a histochemical investigation. *Muscle Nerve* **15,** 63–72.

Jennings, C. G. B. and Burden, S. J. (1993) Development of the neuromuscular synapse. *Curr. Biol.* **3,** 75–81.

Jo, S. A., Zhu, X., Marchionni, M. A., and Burden, S. J. (1995) Neuregulins are concentrated at nerve-muscle synapses and activate acetylcholine receptor gene expression. *Nature* **373,** 158–161.

Johnson, C. D. and Russell, R. L. (1975) A rapid, simple radiometric assay for cholinesterase, suitable for multiple determinations. *Anal. Biochem.* **64,** 229–238.

Jonas, E. A., Snape, A. M., and Sargent, T. D. (1989) Transcriptional regulation of a *Xenopus* embryonic epidermal keratin gene. *Development* **106,** 399–405.

Kadar, T., Silberman, M., Weissman, B. A., and Levy, A. (1990) Age-related changes in the cholinergic components within the central nervous system II. Working memory impairment and its relation to hippocampal muscarinic receptors. *Mech. Ageing Dev.* **55,** 139–149.

Karnovsky, M. J. (1964) The localization of cholinesterase activity in rat cardiac muscle by electron microscopy. *J. Cell Biol.* **23,** 217–232.

Karnovsky, M. J. and Roots, L. (1964) A "direct coloring" thiocholine method for cholinesterases. *J. Histochem. Cytochem.* **12,** 219–221.

Karpel, R., Ben Aziz-Aloya, R., Sternfeld, M., Ehrlich, G., Ginzberg, D., Tarroni, P., Clementi, F., Zakut, H. and Soreq, H. (1994) Expression of three alternative acetylcholinesterase messenger RNAs in human tumor cell lines of different tissue origins. *Exp. Cell Res.* **210,** 268–277.

Karpel, R., Sternfeld, M., Ginzberg, D., Huhl, E., Graessmann, A., and Soreq, H. (1995) Overexpression of alternative human acetylcholinesterase forms modulates process extensions in cultured glioma cells. *J. Neurochem.* **66,** 114–123.

Katz, B. and Miledi, R. (1973) The binding of acetylcholine to receptors and its removal from the synaptic cleft. *J. Physiol. (Lond.)* **231,** 549–574.

Katzman, R. (1986) Alzheimer's disease. *N. Engl. J. Med.* **314,** 964–973.

Kay, B. K. and Peng, H. J. (eds.) (1991) *Methods in Cell Biology,* vol. 36, Academic, San Diego.

Keller, R. (1991) Early embryonic development of *Xenopus laevis*, in *Methods in Cell Biology*, vol. 36 (Kay, B. K. and Peng, H. J., eds.), Academic, San Diego, pp. 61–113.

Keller, R. and Danilchik, M. (1988) Regional expression, pattern and timing of convergence and extension during gastrulation of *Xenopus laevis*. *Development* **103,** 193–209.

Kelly, G. M., Eib, D. W., and Moon, R. T. (1991) Histological preparation of *Xenopus laevis* oocytes and embryos, in *Methods in Cell Biology*, vol. 36 (Kay, B. K. and Peng, H. J., eds.), Academic, San Diego, pp. 389–417.

Kerem, A., Kronman, C., Bar Nun, S., Shafferman, A., and Velan, B. (1993) Interrelations between assembly and secretion of recombinant human acetylcholinesterase. *J. Biol. Chem.* **268,** 180–184.

Kimelman, D., Christian, J. L., and Moon, R. T. (1992) Synergistic principles of development: overlapping patterning systems in *Xenopus* mesoderm induction. *Development* **116,** 1–9.

Kloc, M. and Etkin, L. D. (1995) Two distinct pathways for the localization of RNAs at the vegetal cortex in *Xenopus* oocytes. *Development* **121,** 287–297.

Klymkowsky, M. W. and Hanken, J. (1991) Whole mount staining of *Xenopus* and other vertebrates, in *Methods in Cell Biology*, vol. 36 (Kay, B. K. and Peng, H. J., eds.), Academic, San Diego, pp. 420,421.

Klymkowsky, M. W., Maynell, L. A., and Polson, A. G. (1987) Polar asymmetry in the organization of the cortical cytokeratin system of *Xenopus laevis* oocytes and embryos. *Development* **100,** 543–557.

Knapp, M. J., Knopman, D. S., Solomon, P. R., Pendlebury, W. W., Davis, C. S., and Gracon, S. I. (1994) A 30-week randomized controlled trial of high-dose Tacrine in patients with Alzheimer's disease. *JAMA* **271,** 985–991.

Koelle, G. B. and Friedenwald, J. S. (1949) A histochemical method for localizing cholinesterase activity. *Proc. Soc. Exp. Biol. Med.* **70,** 617–622.

Krejci, E., Coussen, F., Duval, N., Chatel, J. M., Legay, C., Puype, M., Vanderkerckhove, J., Cartaud, J., Bon, S., and Massoulie, J. (1991) Primary structure of a collagenic tail peptide of *Torpedo* acetylcholinesterase: co-expression with catalytic subunit induces the production of collagen-tailed forms in transfected cells. *EMBO J.* **10,** 1285–1293.

Krieg, P. A. and Melton, D. A. (1985) Developmental regulation of a gastrula-specific gene injected into fertilized *Xenopus* eggs. *EMBO J.* **4,** 3463–3471.

Krieg, P. A. and Melton, D. A. (1987) An enhancer responsible for activating transcription at the MBT in *Xenopus* development. *Proc. Natl. Acad. Sci. USA* **84,** 2331–2335.

Kronman, C., Velan, B., Gozes, Y., Leitner, M., Flashner, Y., Lazar, A., Marcus, D., Sery, T., Papier, Y., Grosfeld, H., Cohen, S., and Shafferman, A. (1992) Production and secretion of high levels of recombinant human acetylcholinesterase in cultured cell lines: microheterogeneity of the catalytic subunit. *Gene* **121,** 295–304.

Ku, M. and Melton, D. A. (1993) Xwnt-11: a maternally expressed *Xenopus* wnt gene. *Development* **119,** 1161–1173.

Kullberg, R. W., Lentz, T. L., and Cohen, M. W. (1977) Development of the myotomal neuromuscular junction in *Xenopus laevis*: an electrophysiological and fine-structural study. *Dev. Biol.* **60,** 101–129.

Kullberg, R. W., Mikelberg, F. S., and Cohen, M. W. (1980) Contribution of cholinesterase to developmental decreases in the time course of synaptic potentials at an amphibian neuromuscular junction. *Dev. Biol.* **75,** 255–267.

Lai, C. J., Ekker, S. C., Beachy, P. A., and Moon, R. T. (1995) Patterning of neural ectoderm of *Xenopus laevis* by the amino-terminal product of hedgehog autoproteolytic cleavage. *Development* **121,** 2349–2360.

Lamb, T. M., Knecht, A. K., Smith, W. C., Stachel, S. E., Economides, E., Stahl, N., Yancopolous, G. D., and Harland, R. M. (1993) Neural induction by the secreted polypeptide *noggin*. *Science* **262,** 713–718.

Land, B. R., Harris, W. V., Salpeter, E. E., and Salpeter, M. M. (1984) Diffusion and binding constants for acetylcholine derived from the falling phase of miniature endplate currents. *Proc. Natl. Acad. Sci. USA* **81,** 1594–1598.

Land, B. R., Salpeter, E. E., and Salpeter, M. M. (1980) Acetylcholine receptor site density affects the rising phase of miniature endplate currents. *Proc. Natl. Acad. Sci. USA* **77,** 3736–3740.

Land, B. R., Salpeter, E. E., and Salpeter, M. M. (1981) Kinetic parameters for acetylcholine interaction in intact neuromuscular junctions. *Proc. Natl. Acad. Sci. USA* **78,** 7200–7204.

Landwehrmeyer, B., Probst, A., Palacios, J. M., and Mengod, G. (1993) Expression of acetylcholinesterase messenger RNA in human brain: an in situ hybridization study. *Neuroscience* **57,** 615–634.

Lane, C. D., Marbaix, G., and Gurdon, J. B. (1971) Rabbit haemoglobin synthesis in frog cells: the translation of reticulocyte 9S RNA in frog oocytes. *J. Mol. Biol.* **61,** 73–91.

Lapidot-Lifson, Y., Patinkin, D., Prody, C. A., Ehrlich, G., Seidman, S., Ben Aziz, R., Benseler, F., Eckstein, F., Zakut, H., and Soreq, H. (1992) Cloning and antisense oligodeoxynucleotide inhibition of a human homolog of cdc2 required in hematopoiesis. *Proc. Natl. Acad. Sci. USA* **89,** 579–583.

Lapidot-Lifson, Y., Prody, C. A., Ginzberg, D., Meytes, D., Zakut, H., and Soreq, H. (1989) Co-amplification of human acetylcholinesterase and butyrylcholinesterase in blood cells: correlation with various leukemias and abnormal megakaryocytopoiesis. *Proc. Natl. Acad. Sci. USA* **86,** 4715–4717.

Lappin, R. I. and Rubin, L. L. (1985) Molecular forms of acetylcholinesterase in *Xenopus* muscle. *Dev. Biol.* **110,** 269–274.

Laskowski, M. B., Olson, W. H., and Dettbarn, W. D. (1975) Ultrastructural changes at the motor end-plate produced by an irreversible cholinesterase inhibitor. *Exp. Neurol.* **47,** 290–306.

Lauder, J. M. (1993) Neurotransmitters as growth regulatory factors. *TINS* **16,** 233–240.

Layer, P. G., Weikert, T., and Alber, R. (1993) Cholinesterases regulate neurite growth of chick nerve cells *in vitro* by means of a non-enzymatic mechanism. *Cell. Tissue Res.* **273,** 219–226.

Layer, P. G. and Willbold, E. (1995) Novel functions of cholinesterases in development, physiology and disease. *Prog. Histochem. Cytochem* **29,** 1–92.

Lee, J. E., Hollenberg, S. M., Snider, L., Turner, D. L., Lipnick, N., and Weintraub, H. (1995) Conversion of *Xenopus* ectoderm into neurons by NeuroD, a basic helix-loop-helix protein. *Science* **268,** 836–844.

Legay, C., Bon, S., and Massoulie, J. (1993a) Expression of a cDNA encoding the glycolipid-anchored form of rat acetylcholinesterase. *FEBS Lett.* **315,** 163–166.

Legay, C., Bon, S., Vernier, P., Coussen, F., and Massoulie, J. (1993b) Cloning and expression of a rat acetylcholinesterase subunit: generation of multiple molecular forms and complementarity with a *Torpedo* collagenic subunit. *J. Neurochem.* **60,** 337–346.

Leibham, J. D., Wong, M. W., Cheng, T.-C., Schroeder, S., Weil, P. A., Olson, E. N., and Perry, M. (1994) Binding of TFIID and MEF2 to the TATA element activates transcription of the *Xenopus* MyoDa promoter. *Mol. Cell. Biol.* **14,** 686–689.

Lemaire, P., Garrett, N., and Gurdon, J. B. (1995) Expression cloning of Siamois, a *Xenopus* homeobox gene expressed in dorsal-vegetal cells of blastulae and able to induce a complete secondary axis. *Cell* **81,** 85–94.

Leonard, J. P. and Salpeter, M. M. (1978) Agonist-induced myopathy at the neuromuscular junction is mediated by Calcium. *J. Cell Biol.* **82,** 811–819.

Lester, H. A. (1988) Heterologous expression of excitability proteins: route to more specific drugs? *Science* **241,** 1057–1063.

Lev-Lehman, E., Ginzberg, D., Hornreich, G., Ehrlich, G., Meshorer, A., Eckstein, F., Soreq, H., and Zakut, H. (1994) Antisense inhibition of acetylcholinesterase gene expression causes transient hematopoietic alterations. *Gene Ther.* **1,** 127–135.

Li, Y., Camp, S., Rachinsky, T. L., Getman, D., and Taylor, P. (1991) Gene structure of mammalian acetylcholinesterase. Alternative exons dictate tissue-specific expression. *J. Biol. Chem.* **266,** 23,083–23,090.

Li, Y., Camp, S., and Taylor, P. (1993) Tissue-specific expression and alternative mRNA processing of the mammalian acetylcholinesterase gene. *J. Biol. Chem.* **268,** 5790–5797.

Liao, J., Boschetti, N., Mortensen, V., Jensen, S. P., Koch, C., Norgaard-Pedersen, B., and Brodbeck, U. (1994) Characterization of salt-soluble forms of acetylcholinesterase from bovine brain. *J. Neurochem.* **63,** 1446–1453.

Liao, J., Heider, H., Sun, M. C., and Brodbeck, U. (1992) Different glycosylation in acetylcholinesterases from mammalian brain and erythrocytes. *J. Neurochem.* **58,** 1230–1238.

Liao, J., Heider, H., Sun, M. C., Stieger, S., and Brodbeck, U. (1991) The monoclonal antibody 2G8 is carbohydrate-specific and distinguishes between different forms of vertebrate cholinesterases. *Eur. J. Biochem.* **198,** 59–65.

Liao, J., Mortensen, V., Norgaard-Pedersen, B., Koch, C., and Brodbeck, U. (1993a) Monoclonal antibodies against brain acetylcholinesterases which recognize the subunits bearing the hydrophobic anchor. *Eur. J. Biochem.* **215,** 333–340.

Liao, J., Norgaard-Pedersen, B., and Brodbeck, U. (1993b) Subunit association and glycosylation of acetylcholinesterase from monkey brain. *J. Neurochem.* **61,** 1127–1134.

Lieth, E., Cardasis, C. A., and Fallon, J. R. (1992) Muscle-derived agrin in cultured myotubes: expression in the basal lamina and at induced acetylcholine receptor clusters. *Dev. Biol.* **149,** 41–54.

Livneh, A., Sarova, I., Michaeli, D., Pras, M., Wagner, K., Zakut, H., and Soreq, H. (1988) Antibodies against acetylcholinesterase and low levels of cholinesterases in a patient with an atypical neuromuscular disorder. *Clin. Immunol. Immunopathol.* **48,** 119–131.

Loewenstein, Y., Gnatt, A., Neville, L. F., and Soreq, H. (1993) Chimeric human cholinesterase: identification of interaction sites responsible for recognition of acetyl- or butyrylcholinesterase-specific ligands. *J. Mol. Biol.* **234,** 289–296.

Loewenstein-Lichtenstein, Y., Schwarz, M., Glick, D., Norgaard-Pedersen, B., Zakut, H., and Soreq, H. (1995) Genetic predisposition to adverse consequences of anti-cholinesterases in atypical BCHE carriers. *Nature Med.* **1,** 1082–1085.

Loewi, O. and Navratil, E. (1926) Uber humorale Ubertragbarkeit der herznervenwirkung XI. Uber den mechanismus der baguswirkung von phsostigmin und ergotamin. *Pflügers Arch.* **214,** 689–696.

Lu, B., Greengard, P., and Poo, M.-m. (1992) Exogenous synapsin I promotes functional maturation of developing neuromuscular synapses. *Neuron* **8,** 521–529.

Ludolph, D. C., Nepp, A. W., Mescher, A. L., Malacinski, G. M., Parker, M. A., and Smith, R. C. (1994) Overexpression of XMyoD or Xmyf5 in *Xenopus* embryos induces the formation of enlarged myotomes through recruitment of cells of nonsomitic lineage. *Dev. Biol.* **166,** 18–32.

MacNicol, A. M., Muslin, A. J., and Williams, L. T. (1993) Raf-1 kinase is essential for early *Xenopus* development and mediates the induction of mesoderm by FGF. *Cell* **73,** 571–583.

Mangold, O. (1933) Uber die Induktionsfahigkeit der verscheidenen Bezirke der Nerula von urodelen. *Nautrwissenchaften* **43,** 761–766.

Massoulie, J. and Bon, S. (1982) The molecular forms of cholinesterase and acetylcholinesterase in vertebrates. *Annu. Rev. Neurosci.* **5,** 57–106.

Massoulie, J., Pezzementi, L., Bon, S., Krejci, E., and Vallette, F. M. (1993) Molecular and cellular biology of cholinesterases. *Prog. Neurobiol.* **41,** 31–91.

Martin, P. T., Ettinger, A. J., and Sanes, J. R. (1995) A synaptic localization domain in the synaptic cleft protein laminin β2 (s-Laminin). *Science* **269,** 413–416.

Matthews-Bellinger, J. and Salpeter, M. M. (1978) Distribution of acetylcholine receptors at frog neuromuscular junctions with a discussion of some physiological implications. *J. Physiol.* **279,** 197–213.

Maulet, Y., Camp, S., Gibney, G., Rachinsky, T. L., Ekstrom, T. J., and Taylor, P. (1990) Single gene encodes glycophospholipid-anchored and asymmetric acetylcholinesterase forms: alternative coding exons contain inverted repeat sequences. *Neuron* **4,** 289–301.

Mayser, W., Schloss, P., and Betz, H. (1992) Primary structure and functional expression of a choline transporter expressed in the rat nervous system. *FEBS Lett.* **305,** 31–36.

McCrea, P. D., Brieher, W. M., and Gumbiner, B. M. (1993) Induction of a secondary body axis in *Xenopus* by antibodies to β-catenin. *J. Cell Biol.* **123,** 447–484.

McMahan, U. J., Horton, S. E., Werle, M. J., Honig, L. S., Kroger, S., Ruegg, M. A., and Escher, G. (1992) Agrin isoforms and their role in synaptogenesis. *Curr. Biol.* **4,** 869–874.

McMahan, U. J., Sanes, J. R., and Marshall, L. M. (1978) Cholinesterase is associated with the basal lamina at the neuromuscular junction. *Nature* **271,** 172–174.

McMahon, A. P. and Moon, R. T. (1989) Ectopic expression of the proto-oncogene int-1 in *Xenopus* embryos leads to duplication of the embryonic axis. *Cell* **58,** 1075–1084.

Meflah, K., Bernard, S., and Massoulie, J. (1984) Lectins indicate differences in the carbohydrate composition of the membrane-bound enzymes acetylcholinesterase and 5'-nucleotidase in different cell types. *Biochimie* **66,** 59–69.

Mendolzon, D., Changeux, J. P., and Nghiem, H. O. (1994) Phosphorylation of myogenin in chick myotubes: regulation by electrical activity and protein kinase C. Implications for acetylcholine receptor gene expression. *Biochemistry* **33,** 2568–2575.

Merlie, J. P. and Kornhauser, J. M. (1989) Neural regulation of gene expression by an acetylcholine receptor promoter in muscle of transgenic mice. *Neuron* **2,** 1295–1300.

Merlie, J. P. and Sanes, J. R. (1985) Concentration of acetylcholine receptor mRNA in synaptic regions of adult muscle fibers. *Nature* **317,** 66–68.

Mishina, M., Kurosaki, T., Tobimatsu, T., Morimoto, Y., Noda, M., Yamamoto, T., Terao, M., and Lindstrom, J. (1984) Expression of functional acetylcholine receptor from cloned cDNAs. *Nature* **307,** 604–608.

Mohun, T. J., Taylor, M. V., Garrett, N., and Gurdon, J. B. (1989) The CArG promoter sequence is necessary for muscle-specific transcription of the cardiac actin gene in *Xenopus* embryos. *EMBO J.* **8,** 1153–1161.

Moody-Corbett, F. and Cohen, M. W. (1981) Localization of cholinesterase at sites of high acetylcholine receptor density on embryonic amphibian muscle cells cultured without nerve. *J. Neurosci.* **1,** 596–604.

Morasso, M. L., Jamrich, M., and Sargent, T. D. (1993) The homeodomain gene *Xenopus* distal-less-like (XdII-2) is regulated by a conserved mechanism in amphibian and mammalian epidermis. *Dev. Biol.* **162,** 267–276.

Moreno, S. and Nurse, P. (1990) Substrates for $p34^{cdc2}$: *in vivo* veritas? *Cell* **61,** 549–551.

Mutero, A., Camp, S., and Taylor, P. (1995) Promoter elements of the mouse acetylcholinesterase gene. *J. Biol. Chem.* **270,** 1866–1872.

Mutero, A. and Fournier, D. (1992) Post-translational modifications of *Drosophila* acetylcholinesterase. *In vitro* mutagenesis and expression in *Xenopus* oocytes. *J. Biol. Chem.* **267,** 1695–1700.

Nastuk, M. A. and Fallon, J. R. (1993) The molecular choreography of synapse formation. *TINS* **16,** 72–76.

Navaratnam, D. S., Priddle, J. D., McDonald, B., Esiri, M. M., Robinson, J. E., and Smith, A. D. (1991) Anomalous molecular form of acetylcholinesterase in cerebrospinal fluid in histologically diagnosed Alzheimer's disease. *Lancet* **337,** 447–450.

Nelson, L. (1964) Acetylcholinesterase in bull spermatozoa. *J. Reprod. Fert.* **7,** 65–71.

Neville, L. F., Gnatt, A., Loewenstein, Y., Seidman, S., Ehrlich, G., and Soreq, H. (1992) Intramolecular relationships in cholinesterases revealed by oocyte expression of site-directed and natural variants of human BCHE. *EMBO J.* **11,** 1641–1649.

Neville, L. F., Gnatt, A., Loewenstein, Y., and Soreq, H. (1990a) Aspartate-70 to glycine substitution confers resistance to naturally occurring and synthetic anionic-site ligands on *in-ovo* produced human butyrylcholinesterase. *J. Neurosci. Res.* **27,** 452–460.

Neville, L. F., Gnatt, A., Padan, R., Seidman, S., and Soreq, H. (1990b) Anionic site interactions in human butyrylcholinesterase disrupted by two single point mutations. *J. Biol. Chem.* **265,** 20,735–20,738.

Newport, J. and Kirschner, M. (1982a) A major developmental transition in early *Xenopus* embryos: I. Characterization and timing of cellular changes at the midblastula stage. *Cell* **30,** 675–686.

Newport, J. and Kirschner, M. (1982b) A major developmental transition in early *Xenopus* embryos: II. Control of the onset of transcription. *Cell* **30,** 687–696.

Niehrs, C., Keller, R., Chou, K. W. Y., and De Robertis, E. M. (1993) The homeobox gene goosecoid controls cell migration in *Xenopus* embryos. *Cell* **72,** 491–503.

Nieuwkoop, P. D. (1973) The "organizing center" of the amphibian embryo: its origin, spatial organization, and morphogenic action. *Adv. Morphogen.* **10,** 3–39.

Nieuwkoop, P. D. and Faber, J. (1967) *Normal Table of* Xenopus laevis *(Daudin)*, North-Holland, Amsterdam.

Nieuwkoop, P. D. and Faber, J. (1994) *Normal Table of* Xenopus laevis *(Daudin)*, Garland, New York.

Nishikawa, S. and Sasaki, Y. (1993) Secretion of chondroitin sulfate from embryonic epidermal cells in *Xenopus laevis*. *Histochemistry* **41,** 1373–1381.

Nitkin, R. M., Smith, M. A., Fallon, M. C., Jr., Yao, Y.-M. M., Wallace, B. G., and McMahan, U. J. (1987) Identification of agrin, a synaptic organizing protein from *Torpedo* electric organ. *J. Cell Biol.* **105,** 2471–2478.

Ohlendieck, K., Ervasti, J. M., Matsumura, K., Kahl, S. D., Leveille, C. J., and Campbell, K. P. (1991) Dystrophin-related protein is localized to neuromuscular junctions of adult skeletal muscle. *Neuron* **7,** 499–508.

O'keefe, H. P, Melton, D. A., Ferreiro, B., and Kintner, C. (1991) *In situ* hybridization, in *Methods in Cell Biology*, vol. 36 (Kay, B. K. and Peng, H. J., eds.), Academic, San Diego, pp. 444–463.

Olson, D. J., Christian, J. L., and Moon, R. T. (1991) Effect of Wnt-1 and related proteins on gap junctional communication in *Xenopus* embryos. *Science* **252,** 1171–1176.

Ord, M. G. and Thompson, R. H. (1953) Pseudocholinesterase activity in the central nervous system. *Biochem. J.* **51,** 245–251.

O'Reilly, M. A. J., Smith, J. C., and Cunliffe, V. (1995) Patterning of the mesoderm in *Xenopus*: dose-dependent and synergistic effects of brachyury and pintallavis. *Development* **121,** 1351–1359.

Oron, Y., Gillo, B., and Gershengorn, M. C. (1988) Hemispheric asymmetry of membrane electrical responses in native and mRNA-injected *Xenopus* oocytes. *Proc. Natl. Acad. Sci. USA* **85,** 3820–3824.

Palecek, J., Habrova, V., Nedvidek, J., and Romanovsky, A. (1985) Dynamics of tubulin structures in *Xenopus laevis* oogenesis. *J. Embryol. Exp. Morphol.* **87,** 75–86.

Patinkin, D., Seidman, S., Eckstein, F., Benseler, F., Zakut, H., and Soreq, H. (1990) Manipulations of cholinesterase gene expression modulate murine megakaryocytopoiesis *in vitro. Mol. Cell. Biol.* **10,** 6046–6050.

Patrick, J., Sequela, P., Vernino, S., Amador, M., Luetje, C., and Dani, J. A. (1993) Functional diversity of neuronal nicotinic acetylcholine receptors, in *Progress in Brain Research,* vol. 98 (Cuello, A. C., ed.), Elsevier, Amsterdam, pp. 115–122.

Paul, D. L., Bruzzone, Y. R., Gimlich, R. L., and Goodenough, D. A. (1995) Expression of a dominant negative inhibitor of intercellular communication in the early *Xenopus* embryo causes delamination and extrusion of cells. *Development* **121,** 371–381.

Paulus, J. M., Maigen, J., and Keyhani, E. (1981) Mouse megakaryocytes secrete acetylcholinesterase. *Blood* **58,** 1100–1106.

Pavlath, G. K., Rich, K., Webster, S. G., and Blau, H. M. (1989) Localization of muscle gene products in nuclear domains. *Nature* **337,** 570–573.

Peng, H. B., Baker, L. P., and Chen, Q. (1991) Tissue culture of *Xenopus* neurons and muscle cells as a model for studying synaptic induction, in *Methods in Cell Biology,* vol. 36. (Kay, B. K. and Peng H. B., eds.), Academic, San Diego, pp. 511–526.

Phillips, W. D., Kopta, C., Blount, P., Gardner, P. D., Steinbach, J. H., and Merlie, J. P. (1991) Acetylcholine receptor-rich membrane domains organized in fibroblasts by recombinant 43 kilodalton protein. *Science* **251,** 568–570.

Porter, B. E., Weis, J., and Sanes, J. R. (1995) A motoneuron-selective stop signal in the synaptic protein S-Laminin. *Neuron* **14,** 549–559.

Pressman-Schwartz, S., Aisenthal, L., Zichrini, E., Oberman, F., and Yisraeli, J. R. (1992) A 69-kDa RNA-binding protein from *Xenopus* oocytes recognizes a common motif in two vegetally localized maternal RNAs. *Proc. Natl. Acad. Sci. USA* **89,** 11,895–11,899.

Prody, C. A., Zevin-Sonkin, D., Gnatt, A., Goldberg, O., and Soreq, H. (1987) Isolation and characterization of full-length cDNA clones coding for cholinesterase from fetal human tissues. *Proc. Natl. Acad. Sci. USA* **84,** 3555–3559.

Purcell, L., Gruia-Gray, J., Scanga, J., and Ringuette, M. (1993) Developmental anomalies of Xenopus embryos following microinjection of SPARC antibodies. *J. Exp. Zool.* **265,** 153–164.

Quinn, D. M. (1987) Acetylcholinesterase: enzyme structure, reaction dynamics, and virtual transition states. *Chem. Rev.* **87,** 955–979.

Rachinsky, T. L., Camp, S., Li, Y., Ekstrom, T. J., Newton, M., and Taylor, P. (1990) Molecular cloning of mouse acetylcholinesterase: tissue distribution of alternatively spliced mRNA species. *Neuron* **5,** 317–327.

Radic, Z., Gibney, G., Kawamoto, S., MacPhee Quigley, K., Bongiorno, C., and Taylor, P. (1992) Expression of recombinant acetylcholinesterase in a baculovirus system: kinetic properties of glutamate 199 mutants. *Biochemistry* **31,** 9760–9767.

Rakonczay, Z. and Brimijoin, S. (1988) Biochemistry and pathophysiology of the molecular forms of cholinesterases. *Subcell. Biochem.* **12,** 335–378.

Ralston, E. and Hall, Z. W. (1989) Transfer of a protein encoded by a single nucleus to nearby nuclei in multinucleated mytobes. *Nature* **244,** 1066–1069.

Rama Sastry, B. V. and Sadavongvivad, C. (1979) Cholinergic systems in non-nervous tissues. *Pharmacol. Rev.* **30,** 65–132.

Ramirez, G., Bakat, A., and Fernandez, H. L. (1990) Interaction of asymmetric and globular acetylcholinesterase species with glycosaminoglycans. *J. Neurochem.* **54,** 1761–1768.

Rashbass, J., Taylor, M. V., and Gurdon, J. B. (1992) The DNA-binding protein E12 co-operates with XMyoD in the activation of muscle-specific gene expression in *Xenopus* embryos. *EMBO J.* **11,** 2981–2990.

Reist, N. E., Werle, M. J., and McMahan, U. J. (1992) Agrin released by motor neurons induces the aggregation of acetylcholine receptors at neuromuscular junctions. *Neuron* **8,** 865–868.

Ripoll, D. R., Faerman, C. H., Axelsen, P. H., Silman, I., and Sussman, J. L. (1993) An electrostatic mechanism for substrate guidance down the aromatic gorge of acetylcholinesterase. *Proc. Natl. Acad. Sci. USA* **90,** 5128–5132.

Roberts, W. L., Doctor, B. P., Foster, J. D., and Rosenberry, T. L. (1991) Bovine brain acetylcholinesterase primary sequence involved in intersubunit disulfide linkages. *J. Biol. Chem.* **266,** 7481–7487.

Robertson, R. T. (1993) Acetylcholinesterase of neural development: new tricks for an old dog? *News Physiol. Sci.* **8,** 266–272.

Roelink, H., Augsburger, A., Heemskerk, J., Korzh, V., Norlin, S., Ruiz i Altaba, A., Tanabe, Y., Placzek, M., Edlund, T., and Jessell, E. T. M. (1994) Floor plate and motor neuron induction by vhh-1, a vertebrate homolog of hedgehog expressed by the notochord. *Cell* **76,** 761.

Rossi, S. G. and Rotundo, R. L. (1992) Cell surface acetylcholinesterase molecules on multinucleated myotubes are clustered over the nucleus of origin. *J. Cell Biol.* **119,** 1657–1667.

Rothbacher, U., Laurent, M. N., Blitz, I. L., Watabe, T., Marsh, J. L., and Cho, K. W. (1995) Functional conservation of the Wnt signaling pathway revealed by ectopic expression of *Drosophila* dishevelled in *Xenopus. Dev. Biol.* **170,** 717–721.

Rosenberry, T. (1975) Acetylcholinesterase. *Adv. Enzymol.* **43,** 103–217.

Rosenberry, T. (1979) Quantitative simulation of endplate currents at neuromuscular junctions based on the reaction of acetylcholine with acetylcholine receptor and acetylcholinesterase. *Biophys. J.* **26,** 263–290.

Rotundo, R. L. (1990) Nucleus-specific translation and assembly of acetylcholinesterase in multinucleated muscle cells. *J. Cell Biol.* **110,** 715–719.

Rudy, B. and Iverson, L. E. (eds.) (1992) *Methods in Enzymology,* vol. 207, Academic, New York.

Ruiz i Altaba, A. (1992) Planar and vertical signals in the induction and patterning of the *Xenopus* nervous system. *Development* **115,** 67–80.

Ruiz i Altaba, A. (1994) Pattern formation in the vertebrate neural plate. *TINS* **17,** 233–243.

Ruiz i Altaba, A. and Jessell, T. M. (1992) Pintallavis, a gene expressed in the organizer and midline cells of frog embryos: involvement in the development of the neural axis. *Development* **116,** 81–93.

Ruiz i Altaba, A., Jessell, T. M., and Klar, A. (1993) Ectopic neural expression of a floor plate marker in frog embryos injected with the midline transcription factor pintallavis. *Proc. Natl. Acad. Sci. USA* **90,** 8268–8272.

Ruiz i Altaba, A. and Melton, D. A. (1989) Involvement of the *Xenopus* homeobox gene Xhox3 in pattern formation along the anterior-posterior axis. *Cell* **57,** 317–326.

Rusconi, S. and Schaffner, W. (1981) Transformation of frog embryos with a rabbit β-globin gene. *Proc. Natl. Acad. Sci. USA* **78,** 5051–5055.

Salpeter, M. M. (1967) Electron microscope autoradiography as a quantitative tool in enzyme cytochemistry: I. the distribution of acetylcholinesterase at motor endplates of a vertebrate twitch muscle. *J. Cell Biol.* **32,** 379–389.

Salpeter, M. M. (1987) Vertebrate neuromuscular junctions: general morphology, molecular organization and functional consequences, in *The Vertebrate Neuromuscular Junction* (Salpeter, M. M., ed.), Alan Liss, New York, pp. 1–54.

Salpeter, M. M. and Loring, P. (1985) Nicotinic acetylcholine receptors in vertebrate muscle: properties, distribution and neural control. *Prog. Neurobiol.* **25,** 297–325.

Sambrook, J., Fritsch, E. F., and Maniatis, T. (1989) *Molecular Cloning: A Laboratory Manual,* Cold Spring Harbor Laboratory, Cold Spring Harbor, NY.

Sansom, M. S. P. (1993) Peering down a pore. *Curr. Biol.* **3,** 239–241.

Sargent, T. D. and Mathers, P. H. (1991) Analysis of class II gene regulation, in *Methods in Cell Biology,* vol. 36 (Kay, B. K. and Peng, H. J., eds.), Academic, San Diego, pp. 347–365.

Sassoon, D. A. (1993) Myogenic regulatory factors: dissecting their role and regulation during vertebrate embryogenesis. *Dev. Biol.* **156,** 11–23.

Schaeffer, E., Alder, J., Greengard, P., and Poo, M.-m. (1994) Synapsin IIa accelerates functional development of neuromuscular synapses. *Proc. Natl. Acad. Sci. USA* **91,** 3882–3886.

Schmidt, E. V., Christoph, G., Zeller, R., and Leder, P. (1990) The cytomegalovirus enhancer: a panactive control element in transgenic mice. *Mol. Cell. Biol.* **10,** 4406–4411.

Schonbeck, S., Chrestel, S., and Hohlfeld, R. (1990) *Myasthenia gravis*: prototype of the antireceptor autoimmune diseases. *Int. Rev. Neurobiol.* **32,** 175–200.

Schrag, J. D., Li, Y., Wu, S., and Cygler, M. (1991) Ser-His-Glu triad forms the catalytic site of the lipase from *Geotrichum candidum. Nature* **351,** 761–764.

Schumacher, M., Camp, S., Maulet, Y., Newton, M., MacPhee-Quigley, K., Taylor, S. S., Freidmann, T., and Taylor, P. (1986) Primary structure of *Torpedo* californica acetylcholinesterase deduced from its cDNA sequence. *Nature* **319,** 407–409.

Schutte-Merker, S. and Smith, J. C. (1995) Mesoderm formation in response to brachyury requires FGF signaling. *Curr. Biol.* **5,** 62–67.

Schwarz, M., Glick, D., Loewenstein, Y., and Soreq, H. (1995a) Engineering of human cholinesterases explains and predicts diverse consequences of administration of various drugs and poisons. *Pharmaceut. Ther.* **67,** 283–322.

Schwarz, M., Loewenstein, Y., Glick, D., Liao, J., Norgaard-Pedersen, B., and Soreq, H. (1995b) Successive organophosphate inhibition and oxime reactivation reveals distinct responses of recombinant human cholinesterase variants. *Mol. Br. Res.* **31,** 101–110.

Seidman, S., Ben Aziz-Aloya, R., Timberg, R., Loewenstein, Y., Velan, B., Shafferman, A., Liao, J., Norgaard-Pedersen, B., Brodbeck, U., and Soreq, H. (1994) Overexpressed monomeric human acetylcholinesterase induces subtle ultrastructural modifications in developing neuromuscular junctions of *Xenopus laevis* embryos. *J. Neurochem.* **62,** 1670–1681.

Seidman, S., Sternfeld, M., Ben Aziz-Aloya, R., Timberg, R., Kaufer-Nachum, D., and Soreq, H. (1995) Synaptic and epidermal accumulation of human acetylcholinesterase are encoded by alternative 3'-terminal exons. *Mol. Cell. Biol.* **15,** 2993–3002.

Shafferman, A., Kronman, C., Flashner, Y., Leitner, M., Grosfeld, H., Ordentlich, A., Gozes, Y., Cohen, S., Ariel, N., Barak, D., Harel, M., Silman, I., Sussman, J. L., and Velan, B. (1992a) Mutagenesis of human acetylcholinesterase. Identification of residues involved in catalytic activity and in polypeptide folding. *J. Biol. Chem.* **267,** 17,640–17,648.

Shafferman, A., Ordentlich, A., Barak, D., Kronman, C., Raphael, B., Tamar, B., Naomi, A., Roman, O., and Velan, B. (1994) Electrostatic attraction by surface charge does not contribute to the catalytic efficiency of acetylcholinesterase. *EMBO J.* **13,** 3448–3455.

Shafferman, A., Velan, B., Ordentlich, A., Kronman, C., Grosfeld, H., Leitner, M., Flashner, Y., Cohen, S., Barak, D., and Ariel, N. (1992b) Substrate inhibition of acetylcholinesterase: residues affecting signal transduction from the surface to the catalytic center. *EMBO J.* **11,** 3561–3568.

Shapira, M., Seidman, S., Sternfeld, M., Timberg, R., Kaufer, D., Patrick, J., and Soreq, H. (1994) Transgenic engineering of neuromuscular junctions in *Xenopus laevis* embryos transiently overexpressing key cholinergic proteins. *Proc. Natl. Acad. Sci. USA* **91,** 9072–9076.

Sikorav, J. L., Duval, N., Anselmet, A., Bon, S., Krejci, E., Legay, C., Osterlund, M., Reimund, B., and Massoulie, J. (1988) Complex alternative splicing of acetylcholinesterase transcripts in *Torpedo* electric organ; primary structure of the precursor of the glycolipid-anchored dimeric form. *EMBO J.* **7,** 2983–2993.

Sikorav, J. L., Krejci, E., and Massoulie, J. (1987) cDNA sequences of *Torpedo* marmorata acetylcholinesterase: primary structure of the precursor of a catalytic subunit; existence of multiple 5'-untranslated regions. *EMBO J.* **6,** 1865–1873.

Silman, I. and Futerman, A. H. (1987) Mode of attachment of acetylcholinesterase to the surface membrane. *Eur. J. Biochem.* **170,** 11–22.

Sive, H. (1993) The frog prince-ss: a molecular formula for dorsoventral patterning in *Xenopus*. *Genes Dev.* **7,** 1–12.

Slack, J. M. W. (1994) Inducing factors in *Xenopus* early embryos. *Curr. Biol.* **4,** 116–123.

Slack, J. M. W. and Tannahill, D. (1992) Mechanism of anteroposterior axis specification in vertebrates: lessons from the amphibians. *Development* **114,** 285–302.

Small, D. H., Reed, G., Whitefield, B., and Nurcombe, V. (1995) Cholinergic regulation of neurite outgrowth from isolated chick sympathetic neurons in culture. *J. Neurosci.* **15,** 144–151.

Smit, L. M. E., Hageman, G., Veldman, H., Molenaar, P. C., Oen, B. S., and Jennekens, F. G. I. (1988) A myasthenic syndrome with congenital paucity of secondary clefts: CPSC syndrome. *Muscle Nerve* **11,** 337–348.

Smith, D. (1989) Induction of oocyte maturation: transmembrane signaling events and regulation of the cell cycle. *Development* **107,** 685–699.

Smith, L. D., Xu, W., and Varnold, R. L. (1991) Oogenesis and oocyte isolation, in *Methods in Cell Biology,* vol. 36 (Kay, B. K. and Peng, H. J., eds.), Academic, San Diego, pp. 45–60.

Smith, W. C. and Harland, R. M. (1992) Expression cloning of noggin, a new dorsalizing factor localized to the Spemann organizer in *Xenopus* embryos. *Cell* **70,** 829–840.

Smith, W. C., McKendry, R., Ribisi, S., Jr., and Harland, R. M. (1995) A nodal-related gene defines a physical and functional domain within the Spemann organizer. *Cell* **82,** 37–46.

Smith, L. D., Xu, W., and Varnold, R. L. (1991) Oogenesis and oocyte isolation, in *Methods in Cell Biology,* vol. 36 (Kay, B. K. and Peng, H. J., eds.), Academic, San Diego, pp. 45–60.

Snutch, T. P. (1988) The use of *Xenopus* oocytes to probe synaptic communication. *Trends Neurosci.* **11,** 250–256.

Sokol, S., Christian, J. L., Moon, R. T., and Melton, D. A. (1991) Injected Wnt RNA induces a complete body axis in *Xenopus* embryos. *Cell* **67,** 741–752.

Soreq, H. (1985) The biosynthesis of biologically active proteins in mRNA-microinjected *Xenopus* oocytes. *CRC Crit. Rev. Biochem.* **18,** 199–238.

Soreq, H., Parvari, R., and Silman, I. (1982) Biosynthesis and secretion of catalytically active acetylcholinesterase in *Xenopus* oocytes microinjected with mRNA from rat brain and from *Torpedo* electric organ. *Proc. Natl. Acad. Sci. USA* **79,** 830–834.

Soreq, H., Patinkin, D., Lev-Lehman, E., Grifman, M., Ginzberg, D., Eckstein, F., and Zakut, H. (1994) Antisense oligonucleotide inhibition of acetylcholinesterase gene expression induces progenitor cell expansion and hematopoietic apoptosis *ex vivo*. *Proc. Natl. Acad. Sci. USA* **91,** 7907–7911.

Soreq, H., Ben Aziz, R., Prody, C. A., Seidman, S., Gnatt, A., Neville, L., Lieman-Hurwitz, J., Lev-Lehman, E., Ginzberg, D., Lapidot-Lifson, Y., and Zakut, H. (1990) Molecular cloning and construction of the coding region for human acetylcholinesterase reveals a G + C-rich attenuating structure. *Proc. Natl. Acad. Sci. USA* **87,** 9688–9692.

Soreq, H., Gnatt, A., Loewenstein, Y., and Neville, L. F. (1992) Excavations into the active-site gorge of cholinesterases. *Trends Biochem. Sci.* **17,** 353–358.

Soreq, H. and Seidman, S. (1992) *Xenopus* oocyte microinjection: from gene to protein, in *Methods in Enzymology,* vol. 207 (Rudy, B. and Iverson, L. E., eds.), Academic, New York, pp. 225–265.

Soreq, H., Seidman, S., Dreyfus, P. A., Zevin-Sonkin, D., and Zakut, H. (1989) Expression and tissue-specific assembly of human butyrylcholine esterase in microinjected *Xenopus laevis* oocytes. *J. Biol. Chem.* **264,** 10,608–10,613.

Soreq, H. and Zakut, H. (1990a) Amplification of butyrylcholinesterase and acetylcholinesterase genes in normal and tumor tissues: putative relationship to organophosphorous poisoning. *Pharm. Res.* **7,** 1–7.

Soreq, H. and Zakut, H. (1990b) *Cholinesterase Genes: Multilevelled Regulation. Monographs in Human Genetics* (Sparkes, R. S., ed.), Karger, Basel, Switzerland.

Soreq, H. and Zakut, H. (1993) *Human Cholinesterases and Anticholinesterases,* Academic, San Diego.

Soreq, H., Zevin-Sonkin, D., and Razon, N. (1984) Expression of cholinesterase gene(s) in human brain tissues: translational evidence for multiple mRNA species. *EMBO J.* **3,** 1371–1375.

Spemann, H. and Mangold, H. (1924) Uber induction von embryonalagen durch implantation artfremder organis atoren. *Wilhelm Roux's Arch. Entwicklungsmech Org.* **100,** 599–638.

Steinbesser, H., De Robertis, E. M., Ku, M., Kessler, D. S., and Melton, D. A. (1993) *Xenopus* axis formation and induction of goosecoid by injected Xwnt and activin mRNAs. *Development* **118,** 499–507.

Sussman, J. L., Harel, M., Frolow, F., Oefner, C., Goldman, A., Toker, L., and Silman, I. (1991) Atomic structure of acetylcholinesterase from *Torpedo californica*: a prototypic acetylcholine-binding protein. *Science* **253,** 872–879.

Taira, M., Otani, H., Staint-Jeannet, J. P., and Dawid, I. B. (1995) Role of the LIM class homeodomain protein Xlim-1 in neural and muscle induction by the Spemann organizer in *Xenopus. Nature* **372,** 677–679.

Tannahill, D., Bray, S., and Harris, W. A. (1995) A *Drosophila* E (spl) gene is "neurogenic" in *Xenopus*: a green fluorescent protein study. *Dev. Biol.* **168,** 694–697.

Taylor, P. and Radic, Z. (1994) The cholinesterases: from genes to proteins. *Annu. Rev. Pharmacol. Toxicol.* **34,** 281–320.

Thomsen, G. and Melton, D. A. (1993) Processed Vg1 protein is an axial mesoderm inducer in *Xenopus. Cell* **74,** 433–441.

Thomsen, G., Woolf, T., Whitman, M., Sokol, S., Vaughan, J., Vale, W., and Melton, D. A. (1990) Activins are expressed early in *Xenopus* embryogenesis and can induce axial mesoderm and anterior structures. *Cell* **63,** 485–493.

Toutant, J. P., Roberts, W. L., Murray, N. R., and Rosenberry, T. L. (1989) Conversion of human erythrocyte acetylcholinesterase from an amphiphilic to a hydrophilic form by phosphatidylinositol-specific phospholipase C and serum phospholipase D. *Eur. J. Biochem.* **180,** 503–508.

Tuschl, T., Gohlke, C., Jovin, T. M., Westhof, E., and Eskstein, F. (1994) A three dimensional model for the hammerhead ribozyme based on fluorescence measurements. *Science* **266,** 785–789.

Ullrich, B., Ushkaryov, Y. A., and Sudhof, T. C. (1995) Cartography of neurexins: more than 1000 isoforms generated by alternative splicing and expressed in distinct subsets of neurons. *Neuron* **14,** 497–507.

Valenzuela, D. M., Economides, A. N., Rojas, E., Lamb, T. M., Nunez, L., Jones, P., Lp, N. Y., Espinosa, R., III, Brannan, C. L., and Gilbert, D. J. (1995) Identification of mammalian noggin and its expression in the adult nervous system. *J. Neurosci.* **15,** 6077–6084.

Valtorta, F., Lezzi, N., Benfenati, F., Lu, B., Poo, M.-m., and Greengard, P. (1995) Accelerated structural maturation induced by synapsin I at developing neuromuscular synapses of *Xenopus laevis*. *Eur. J. Neurosci.* **7,** 261–271.

Velan, B., Grosfeld, H., Kronman, C., Leitner, M., Gozes, Y., Lazar, A., Flashner, Y., Marcus, D., Cohen, S., and Shafferman, A. (1991a) The effect of elimination of intersubunit disulfide bonds on the activity, assembly, and secretion of recombinant human acetylcholinesterase. Expression of acetylcholinesterase Cys-580—Ala mutant. *J. Biol. Chem.* **266,** 23,977–23,984.

Velan, B., Kronman, C., Grosfeld, H., Leitner, M., Gozes, Y., Flashner, Y., Sery, T., Cohen, S., Ben Aziz, R., Seidman, S., Shafferman, A., and Soreq, H. (1991b) Recombinant human acetylcholinesterase is secreted from transiently transfected 293 cells as a soluble globular enzyme. *Cell. Mol. Neurobiol.* **11,** 143–156.

Velan, B., Kronman, C., Ordentlich, A., Flashner, Y., Leitner, M., Cohen, S., and Shafferman, A. (1993) N-glycosylation of human acetylcholinesterase: effects on activity, stability and biosynthesis. *Biochem. J.* **296,** 649–656.

Vize, P. D., Melton, D. A., Hemmati-Brivanlou, A., and Harland, R. M. (1991) Assays for gene function in developing *Xenopus* embryos, in *Methods in Cell Biology,* vol. 36 (Kay, B. K. and Peng, H. J., eds.), Academic, San Diego, pp. 367–387.

Wallace, B. G. (1991) The mechanism of agrin-induced acetylcholine receptor aggregation. *Philos. Trans. R. Soc. Lond. Biol.* **331,** 273–280.

Wallace, B. G. (1994) Staurosporine inhibits agrin-induced acetylcholine receptor phosphorylation and aggregation. *J. Cell Biol.* **125,** 661–668.

Wathney, J. C., Nass, M. M., and Lester, H. A. (1979) Numerical reconstruction of the quantal event at nicotinic synapses. *Biophys. J.* **27,** 145–164.

Weeks, D. L. and Melton, D. A. (1987) A maternal mRNA localized to the vegetal hemisphere in *Xenopus* eggs codes for a growth factor related to TGF-beta. *Cell* **51,** 861–867.

Wilson, C., Cross, G. S., and Woodland, H. R. (1986) Tissue-specific expression of actin genes injected into *Xenopus* embryos. *Cell* **47,** 589–599.

Witta, S. E., Agarwal, V. R., and Sato, S. M. (1995) XIPOU 2, a noggin-inducible gene, has direct neuralizing activity. *Development* **121,** 721–730.

Wokke, J. H. J., Jennekens, F. G. I., Molenaar, P. C., Van den Oord, C. J. M., Oen, B. S., and Busch, H. F. M. (1989) Congenital paucity of secondary synaptic clefts (CPSC) syndrome in 2 adult siblings. *Neurology* **38,** 648–654.

Wright, C. V. E., Cho, K. W. Y., Hardwicke, J., Collins, R. H., and De Robertis, E. M. (1989) Interference with the function of a homeobox gene in *Xenopus* embryos produces malformations of the anterior spinal cord. *Cell* **59,** 81–93.

Wu, D. and Hersh, L. B. (1994) Choline acetyltransferase: celebrating its fiftieth year. *J. Neurochem.* **62,** 1653–1663.

Wu, M. and Gerhart, J. (1991) Raising *Xenopus* in the laboratory, in *Methods in Cell Biology,* vol. 36 (Kay, B. K. and Peng, H. J., eds.), Academic, San Diego, pp. 3–18.

Yisraeli, J. K., Sokol, S., and Melton, D. A. (1990) A two-step model for the localization of maternal mRNA in *Xenopus* oocytes: involvement of microtubules and microfilaments in the translocation and anchoring of Vg1 mRNA. *Development* **108,** 289–298.

Zakut, H., Even, L., Birkenfeld, S., Malinger, G., Zisling, R., and Soreq, H. (1988) Modified properties of serum cholinesterases in primary carcinomas. *Cancer* **61,** 727–737.

Zakut, H., Ehrlich, G., Ayalon, A., Prody, C. A., Malinger, G., Seidman, S., Ginzberg, D., Kehlenbach, R., and Soreq, H. (1990) Acetylcholinesterase and butyrylcholinesterase genes coamplify in primary ovarian carcinomas. *J. Clin. Invest.* **86,** 900–908.

Zakut, H., Matzkel, A., Schejter, E., Avni, A., and Soreq, H. (1985) Polymorphism of acetylcholinesterase in discrete regions of the developing human fetal brain. *J. Neurochem.* **45,** 382–389.

Zheng, J. Q., Felder, M., Connor, J. A., and Poo, M.-m. (1994) Turning of nerve growth cones induced by neurotransmitters. *Nature* **368,** 140–144.

Zon, L. I., Mather, C., Burgess, S., Bolce, M. E., Harland, R. M., and Orkin, S. H. (1991) Expression of GATA-binding proteins during embryonic development in *Xenopus laevis. Proc. Natl. Acad. Sci. USA* **88,** 10,642–10,646.

Index

GPSR Compliance

The European Union's (EU) General Product Safety Regulation (GPSR) is a set of rules that requires consumer products to be safe and our obligations to ensure this.

If you have any concerns about our products, you can contact us on ProductSafety@springernature.com

In case Publisher is established outside the EU, the EU authorized representative is:

Springer Nature Customer Service Center GmbH
Europaplatz 3
69115 Heidelberg, Germany

Kolibri 360 GmbH
Ferdinand-Jühlke-Straße 7
99095 Erfurt, Deutschland
produktsicherheit@kolibri360.de